Arduino for Ham Radio

A Radio Amateur's Guide to Open Source Electronics and Microcontroller Projects

Glen Popiel, KW5GP

Cover Design
Sue Fagan, KB1OKW

Production
Shelly Bloom, WB1ENT
Jodi Morin, KA1JPA
David Pingree, N1NAS

Published by  ARRL 100 YEARS

Copyright © 2014 by
The American Radio Relay League, Inc.

Copyright secured under the Pan-American Convention

All rights reserved. No part of this work may be reproduced in any form except by written permission of the publisher. All rights of translation are reserved.

Printed in the USA

Quedan reservados todos los derechos

ISBN: 978-1-62595-016-1

First Edition

We strive to produce books without errors. Sometimes mistakes do occur, however. When we become aware of problems in our books (other than obvious typographical errors), we post corrections on the ARRL website. If you think you have found an error, please check **www.arrl.org/arduino** for corrections. If you don't find a correction there, please let us know by sending e-mail to **pubsfdbk@arrl.org**.

Contents

Foreword
Acknowledgements
About the Author
About This Book
How This Book is Organized
Introduction

1 Introduction to the Arduino
2 Arduino Boards and Variants
3 Arduino Shields, Modules, and Devices
4 Arduino I/O Methods
5 Arduino Development Environment
6 Arduino Development Station
7 Random Code Practice Generator
8 CW Beacon and Foxhunt Keyer
9 Fan Speed Controller
10 Digital Compass
11 Weather Station
12 RF Probe with LED Bar Graph
13 Solar Battery Charge Monitor
14 On-Air Indicator
15 Talking SWR Meter
16 Talking GPS/UTC Time/Grid Square Indicator
17 Iambic Keyer
18 Waveform Generator
19 PS/2 CW Keyboard
20 Field Day Satellite Tracker
21 Azimuth/Elevation Rotator Controller
22 CW Decoder
23 Lightning Detector
24 CDE/Hy-Gain Rotator Controller
25 Modified CDE/Hy-Gain Rotator Controller
26 In Conclusion

Appendix A — Sketches and Libraries
Appendix B — Design and Schematic Tools
Appendix C — Vendor Links and References

Foreword

Homebrewing — "do it yourself" in today's terminology — has been a part of Amateur Radio since the early days. Hams build equipment and antennas, integrate individual station pieces into complete systems, and find new ways to use computers to make operating more efficient or enjoyable.

These days, hams are exploring the world of microcontrollers to create new and exciting ham radio station gear. In this book, author Glen Popiel, KW5GP, describes the popular and inexpensive Arduino microcontroller family and shows how to use these powerful yet inexpensive devices with additional modules, accessory boards and components to create a wide variety of interesting projects. Later chapters describe practical applications that range from a simple digital compass, to CW operating accessories, to test equipment and sophisticated rotator controller/computer interfaces. There's a little something for everyone here. You may get started right away and use the projects as-is, or customize them if you are so inclined.

Arduino hardware and software are all Open Source, which means that they are well documented. Software, schematics and other information is freely available. Support is available online in the form of tutorials and Arduino user groups, so you won't have to go it alone. Part of the appeal of using the Arduino and related modules is the spirit of sharing in the Open Source community. Chances are good that if you are interested in modifying one of these projects to suit your needs, someone has done a similar project and shared the results online.

We hope you'll be inspired to expand your horizons by learning about the Arduino and trying some of the projects in this book.

David Sumner, K1ZZ
Chief Executive Officer
Newington, Connecticut
May 2014

Acknowledgements

To my Dad, who taught me that learning to do things yourself is the best way to learn.

I would like to thank my friend, Tim Billingsley, KD5CKP, for introducing me to the Arduino, acting as my sounding board as the concept for this book came to be, and for trying to keep me sane throughout this whole process. I would also like to thank Craig Behrens, NM4T, for his knowledge, guidance and support.

There are so many others who helped make this book happen, and I apologize in advance to anyone I may have omitted. Thanks to the Olive Branch Mississippi and Helena Arkansas Amateur Radio Clubs, along with Chip Isaacks, W5WWI, for their support and encouragement. I would also like to thank ARRL Publications Manager, Steve Ford, WB8IMY, my editor, Mark Wilson, K1RO, and the staff at ARRL for allowing me the opportunity to work with them.

Many of the images shown in this book depict the use of the Fritzing software tool. See the Fritzing Open Source Hardware Initiative at **www.fritzing.com**.

And a special thanks to the Open Source Community. It is their spirit of sharing knowledge and the fruits of their labors with the world that has made the Arduino the wonderful development platform that it is. Thank you.

About the Author

Glen Popiel is a network engineer and technology consultant for Ciber, Inc. and the Mississippi Department of Education, specializing in Open Source technology solutions. First published in *Kilobaud Microcomputing* in 1979 for circuits he designed for the RCA 1802 microprocessor, he continues to work with microcontrollers and their uses in Amateur Radio and has written numerous articles on computers and Amateur Radio.

Always taking things apart (and sometimes even getting them to work afterward), he discovered electronics in high school and has never looked back. As a teenager, he had one of the first true "home computers," a Digital

Equipment (DEC) PDP-8 minicomputer (complete with state-of-the-art Model 35 Teletype) in his bedroom that he and his friends salvaged from the scrap heap. Over his 40+ year career, he has worked for various aerospace and computer manufacturers on radio and military turbojet research data acquisition and control systems.

Since discovering the Arduino several years ago, he has developed a passion for this powerful, inexpensive microcontroller and has given a number of seminars and hamfest forums on the subject of the Arduino and Open Source. He is a member of the Olive Branch Amateur Radio Club (OBARC), QRP Amateur Radio Club International (QRP-ARCI), and the QRP SkunkWerks, a design team of fellow hams and Arduino enthusiasts who have succeeded in getting the JT65-HF digital mode working natively on the TEN-TEC Rebel, a CW-only (so they thought) QRP transceiver.

Glen is also a former cat show judge and has exhibited Maine Coon cats all over the country, with the highlight being a Best in Show at Madison Square Garden in 1989. He now lives in Southaven, Mississippi, where he continues to create fun and exciting new Arduino projects for Amateur Radio with his trusty Maine Coon sidekick, Mysti.

About This Book

Welcome fellow hams and Arduino enthusiasts.

As a ham "homebrewer" and Arduino "Maker," I wanted to create a group of Arduino projects for the Amateur Radio community in general. Going beyond the typical "blinking lights" programs, assembled here you will find a diverse collection of Arduino ham radio-related projects that can be built in a few short days.

My goal was to provide not just a group of ham-related projects for the Arduino for you to copy and build, but to encourage you to take what is presented here and expand on each project, adding your own personal touches to the finished product. To help you on this journey, this book starts by building a solid foundation with an introduction to the various Arduino boards and add-on components I have found to be most useful in ham radio applications. Each new module or component is described in detail, to assist you in choosing the right pieces for your own projects.

This book assumes a basic working knowledge of electronic components and construction techniques. You don't have to be a master, but you should feel comfortable soldering and building projects. If you're new to electronics, ARRL has some excellent books to help you along the way. For starters, there is *Electronics for Dummies* by Cathleen Shamieh and Gordon McComb (ISBN 978-0470286975, ARRL order no. 0196). Another excellent starting point is *Understanding Basic Electronics* by Walter Banzhaf, WB1ANE (ISBN 978-087259-082-3, ARRL order no. 0823), and of course, there is the timeless *ARRL Handbook*, which is published annually. Every ham should have a copy of the *ARRL Handbook* on their bookshelf. I find myself constantly re-reading mine and learning something new every time. These books and other useful resources are available from Amateur Radio dealers or **www.arrl.org/shop**.

You will also need a working knowledge of the Arduino and the Arduino Integrated Development Environment (IDE). There are several excellent introductory books and tutorials you may find helpful. My personal favorites include *Beginning Arduino* by Michael McRoberts (ISBN 978-1430232407) and *Arduino Cookbook* by Michael Margolis (ISBN 978-1449313876). There are also some outstanding Arduino tutorials online at **www.arduino.cc** and **www.learn.adafruit.com**.

You don't have to be a ham radio operator to build and use the projects in the book, but I do strongly encourage you to become a ham if you are not. You don't know what you're missing. There is something for everyone in the ham radio community, and there is absolutely no reason for you to go it alone as you start out on these projects. Find a local club at **www.arrl.org/find-a-club** and attend a meeting or two. You will not find a friendlier, more helpful bunch of people anywhere, and odds are you will meet other Arduino enthusiasts with whom you can collaborate on your own Arduino projects. For more information on how to become a radio amateur, check out the *ARRL Ham Radio License Manual* (ISBN 978-1-62595-013-0, ARRL order no. 0222) and *Ham Radio for Dummies* (ISBN 978-1-118-59211-3, ARRL order no. 0502), both by H. Ward Silver, NØAX.

How This Book is Organized

This book is designed to introduce the Arduino and how it can be applied to ham radio. The projects presented begin with simple designs and concepts, gradually increasing in complexity and functionality. Each new component or programming technique is explained in detail as it appears in a project.

Chapter 1, *Introduction to the Arduino*, introduces the Arduino and its history, and provides a basic understanding of the concepts of Open Source and the various Open Source licenses.

Chapter 2, *Arduino Boards and Variants*, describes the types of Arduino and Arduino-compatible boards commonly used in ham radio projects.

Chapter 3, *Arduino Shields, Modules, and Devices*, covers various boards and components that can be used to interface with the Arduino, allowing it to sense and communicate with the outside world, with an emphasis on components suited for ham radio projects.

Chapter 4, *Arduino I/O Methods*, discusses in detail the I/O capabilities of the Arduino, which method is best for communicating with the various shields and components, and how best to implement each I/O method.

Chapter 5, *Arduino Development Environment*, introduces the Arduino IDE, writing sketches (programs), using program libraries, and troubleshooting methods.

Chapter 6, *Arduino Development Station*, discusses how best to build a work area to develop projects with the Arduino, including design, breadboarding, prototyping, and construction techniques.

Chapter 7, *Random Code Practice Generator*, introduces a simple Morse Code trainer project to become familiar with the Arduino.

Chapter 8, *CW Beacon and Foxhunt Keyer*, describes another easy project that introduces how to interface with radios.

Chapter 9, *Fan Speed Controller*, shows how to sense temperature and control the speed of a fan to maintain a constant temperature.

Chapter 10, *Digital Compass*, covers how to build a simple digital compass using a 3-axis compass module.

Chapter 11, *Weather Station*, shows how to interface to barometric pressure, humidity, and temperature sensors.

Chapter 12, *RF Probe with LED Bar Graph* measures the relative strength of an RF signal.

Chapter 13, *Solar Battery Charge Monitor*, demonstrates how to use an Arduino to measure solar cell output, battery voltage, and charging current.

Chapter 14, *On-Air Indicator*, demonstrates how to use an Arduino to sense RF and light an On-Air indicator with a programmable delay.

Chapter 15, *Talking SWR Meter*, demonstrates how to use sensors and analog-to-digital conversion methods to determine the standing wave ratio (SWR) of an antenna system. It uses a text-to-speech module to convert the output into speech.

Chapter 16, *Talking GPS/UTC Time/Grid Square Indicator*, shows how to interface a GPS module to the Arduino, calculate and display the Maidenhead Grid Locator, and output the results using a text-to-speech module.

Chapter 17, *Iambic Keyer,* shows how to create a CW keyer using the Arduino.

Chapter 18, *Waveform Generator*, shows how to use a programmable direct digital synthesis (DDS) module to generate sine, square, and triangle waves.

Chapter 19, *PS/2 CW Keyboard*, demonstrates how to send Morse code using a standard PC keyboard.

Chapter 20, *Field Day Satellite Tracker*, describes how to build a model satellite tracker interfaced with PC software such as *Ham Radio Deluxe* and *SatPC32* to track satellites for portable events such as Field Day.

Chapter 21, *Azimuth/Elevation Rotator Controller*, describes how to build an interface to control the Yaesu G5400/5500 satellite rotator controller, sense the position of the antenna, and allow PC software such as *Ham Radio Deluxe* to rotate your antennas and track satellites automatically.

Chapter 22, *CW Decoder*, demonstrates how to interface audio from a receiver and decode incoming CW signals into text on a display.

Chapter 23, *Lightning Detector*, demonstrates how to sense lightning, and calculate the distance and intensity.

Chapter 24, *CDE/Hy-Gain Rotator Controller*, describes how to build an antenna rotator controller, sense the position of the antenna, and drive relays to control your antenna rotator motor.

Chapter 25, *Modified CDE/Hy-Gain Rotator Controller*, describes how to modify a CDE/Hy-Gain HAM series rotator controller, sense the position of the antenna, and allow PC software such as *Ham Radio Deluxe* to automatically control your antenna position.

Chapter 26, *In Conclusion*, discusses projects not included in the book to provide concepts and ideas for other projects to encourage going beyond the scope of the book.

Appendix A, *Sketches and Libraries*, includes a complete summary of the program listings (sketches) and libraries needed for each project and provides information about how to find these listings.

Appendix B, *Design and Schematic Tools*, discusses the *Fritzing* and Cadsoft *EaglePCB* software packages used to develop the projects in this book.

Appendix C, *Vendor Links and References*, provides a listing of Arduino-related resources for components, tutorials, and other relevant information.

Introduction

Building equipment, or homebrewing as hams prefer to call it, has always been a cornerstone of Amateur Radio. In the early days, the only way to get on the air was to build it yourself or bribe a friend to help you build a radio. Later, companies such as Heathkit flourished with kits of all descriptions. I remember eagerly waiting for each new Heathkit catalog to see what new and exciting kit I could dream about building next.

I built many of those kits over the years. In fact, my very first radio as a Novice class licensee was the venerable Heathkit HW-16 transceiver. I didn't build that one, but it had been modified by its builder. That custom-built kit radio inspired me to build Heathkit radios for myself. Not satisfied with what I had, modifying the kits I had just finished building was the next order of business. I'll never forget the day when I finished installing a receive preamplifier in my trusty Heathkit HW-101 HF transceiver so I could hear the newest Amateur Radio satellite, AMSAT-OSCAR 7, on 10 meters. It would not surprise me one bit if that old HW-101 is still out there somewhere, with its current owner wondering about the function of that little switch on the front panel.

With the advent of the modern era, computers, and the complexity of a typical ham shack, homebrewing has taken a back seat to commercially available new rigs and accessories. Personal computers are now an integral part of many ham shacks. Why would anyone want to take the time to build something, or spend months writing a program, when a tried-and-true version is just a credit card and a mouse click or two away? And so, the art of kit building and homebrewing began to fade into the background.

Enter the Arduino, a small, inexpensive, easy-to-program microcontroller. A whole new world is now opened up to the homebrewer. Those once-complex projects that would take too much time, money, and knowledge to complete are now just pennies and a few small steps away.

The Arduino is not without a following of its own. With its Open Source model, it comes with a whole community of developers and builders who call themselves "Makers." Now you have a simple development platform and a whole world of developers sharing the fruits of their labor with everyone…for free.

It was only a matter of time before the lines between the two groups of enthusiasts began to blur. Like that brave enterprising soul that first combined peanut butter with jelly, a synergy formed between ham radio and the

Peanut Butter and Jelly

As it turns outs out, jelly was not the first companion for peanut butter. Originally created by Dr. Ambrose W. Straub in 1880 as a source of nutrition for his patients with bad teeth, peanut butter rapidly grew in popularity, and was a hit at the 1904 World's Fair in St Louis. At the turn of the century, it was considered a delicacy, often served with pimento, nasturtium (edible flowers), cheese, celery, and watercress. It wasn't until 1901 that the first recipe for peanut butter and jelly was published.

During World War II, both peanut butter and jelly were on the US military ration menu and it is said the soldiers used jelly on their peanut butter to make it more palatable. Once the soldiers returned home from the war, peanut butter and jelly became a staple for lunch everywhere.

Arduino. (Stop and think about it — someone had to be first, with no idea what peanut butter and jelly would taste like. Odds are jelly was not their first test ingredient.) The interest in homebrewing is back on the rise, and there are now all sorts of new and exciting projects for ham radio using the Arduino and other microcontrollers.

Presented here is a collection of ham-related projects for you to build and expand upon. Never being one to do the Arduino "blinking light" thing, each project was chosen for its usefulness and functionality in ham activities. By design, the projects are complete and usable as they are, but they leave plenty of room for enhancement and expansion. That's where you come in. Your mission, should you decide to accept it, is to take these projects, add to them, and make them better and more powerful than they are. That's where the real fun is in Open Source development — taking someone else's project and making it better, tweaking it to your own personal needs and gaining the satisfaction in knowing that you did it yourself.

And please, don't forget the cardinal rule of Open Source. When you do add your own personal touches and enhance a project beyond what it is, give back to the Open Source community. Who knows, there may be someone out there looking for exactly what you have created. It's that spirit of sharing knowledge and designs that make the Open Source world what it is, a truly unique place to build and share wonderful new toys.

73,
Glen Popiel, KW5GP
kw5gp@arrl.net
May 2014

About the ARRL

The seed for Amateur Radio was planted in the 1890s, when Guglielmo Marconi began his experiments in wireless telegraphy. Soon he was joined by dozens, then hundreds, of others who were enthusiastic about sending and receiving messages through the air—some with a commercial interest, but others solely out of a love for this new communications medium. The United States government began licensing Amateur Radio operators in 1912.

By 1914, there were thousands of Amateur Radio operators—hams—in the United States. Hiram Percy Maxim, a leading Hartford, Connecticut inventor and industrialist, saw the need for an organization to band together this fledgling group of radio experimenters. In May 1914 he founded the American Radio Relay League (ARRL) to meet that need.

Today ARRL, with approximately 155,000 members, is the largest organization of radio amateurs in the United States. The ARRL is a not-for-profit organization that:
- promotes interest in Amateur Radio communications and experimentation
- represents US radio amateurs in legislative matters, and
- maintains fraternalism and a high standard of conduct among Amateur Radio operators.

At ARRL headquarters in the Hartford suburb of Newington, the staff helps serve the needs of members. ARRL is also International Secretariat for the International Amateur Radio Union, which is made up of similar societies in 150 countries around the world.

ARRL publishes the monthly journal *QST* and an interactive digital version of *QST*, as well as newsletters and many publications covering all aspects of Amateur Radio. Its headquarters station, W1AW, transmits bulletins of interest to radio amateurs and Morse code practice sessions. The ARRL also coordinates an extensive field organization, which includes volunteers who provide technical information and other support services for radio amateurs as well as communications for public-service activities. In addition, ARRL represents US amateurs with the Federal Communications Commission and other government agencies in the US and abroad.

Membership in ARRL means much more than receiving *QST* each month. In addition to the services already described, ARRL offers membership services on a personal level, such as the Technical Information Service—where members can get answers by phone, email or the ARRL website, to all their technical and operating questions.

Full ARRL membership (available only to licensed radio amateurs) gives you a voice in how the affairs of the organization are governed. ARRL policy is set by a Board of Directors (one from each of 15 Divisions). Each year, one-third of the ARRL Board of Directors stands for election by the full members they represent. The day-to-day operation of ARRL HQ is managed by an Executive Vice President and his staff.

No matter what aspect of Amateur Radio attracts you, ARRL membership is relevant and important. There would be no Amateur Radio as we know it today were it not for the ARRL. We would be happy to welcome you as a member! (An Amateur Radio license is not required for Associate Membership.) For more information about ARRL and answers to any questions you may have about Amateur Radio, write or call:

ARRL—the national association for Amateur Radio®
225 Main Street
Newington CT 06111-1494
Voice: 860-594-0200
Fax: 860-594-0259
E-mail: **hq@arrl.org**
Internet: **www.arrl.org**

Prospective new amateurs call (toll-free):
800-32-NEW HAM (800-326-3942)
You can also contact us via e-mail at **newham@arrl.org**
r check out the ARRL website at **www.arrl.org**

CHAPTER 1

Introduction to the Arduino

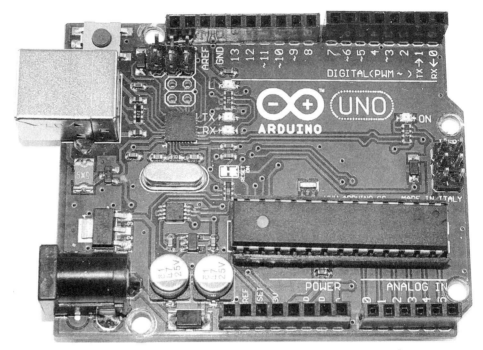

The Arduino Uno.

The Arduino has become wildly popular among the hobbyist community. In 2011, there were an estimated 300,000 Arduino boards in use, not counting the many "clone" boards produced under the Arduino's unique Open Source licensing model. With its standalone single-board design, the Arduino can interface with a wide variety of sensors and controls easily and inexpensively. Based on the Atmel series of microcontrollers, the Arduino, with its onboard digital and analog I/O (input/output), is an easy and inexpensive way to build extremely versatile electronic projects.

Released under the Open Source Creative Commons Attribution Share-Alike license, the Arduino is totally Open Source, as described later in this chapter. From the board designs and schematic files, to the Arduino programs (known as "sketches") and libraries, everything is Open Source. You are free to do whatever you desire, as long as you properly credit the authors in your work and share any changes you make to the existing code and libraries. For the most part, this means that everything about the Arduino is either free or very low cost.

One of the main benefits of Open Source is that you have a whole community of hobbyists developing and sharing their projects freely. This can save you many hours of work if someone is working on projects similar to yours. You can freely integrate their libraries and code into your project, turning what could have been a months-long programming ordeal into a much shorter, more enjoyable path to a finished project

Along with the Arduino board itself, there is a vast selection of components and modules designed to interface with the Arduino. These devices use the various device communication protocols such as the Serial Peripheral Interface (SPI) and Inter-Integrated Circuit (I^2C) already built into the Arduino, allowing simple connections to the Arduino using only a few wires. Now you can create complex projects without having to dig through datasheets and solder for months as you had to in days gone by. For example, the Lightning Detector project presented later in this book needs only 11 wires to connect the lightning detector module and the Nokia LCD display to the Arduino. Since the libraries to communicate with these modules already existed, all that I had to do was include the libraries in the project and get right down to the brass tacks of what I wanted the project to be.

The Hardware

Although there are now numerous variations on the Arduino, the most common Arduino, the Uno consists of an Atmel ATmega328 8-bit microcontroller with a clock speed of 16 MHz. The ATmega328 has 32 KB of flash memory, 2 KB of static RAM (SRAM), and 1 KB of electrically erasable programmable read-only memory (EEPROM) onboard. The Arduino has 14 digital I/O pins. Six of these pins can also do pulse width modulation (PWM), and six 10-bit analog inputs can also be used for digital I/O pins. Two of the digital pins also directly support external hardware interrupts, and all 24 I/O pins support pin state change interrupts, allowing external hardware control of program execution.

Typically powered via the USB programming port, with its low current drain and onboard power regulator the Arduino is ideally suited for battery powered projects. The Arduino supports multiple communication protocols, including standard Serial, Serial Peripheral Interface (SPI), Two-Wire (also known as Inter-Integrated Circuit or I^2C), and 1-Wire. Designed for expandability, the Arduino I/O and power connections are brought out to a series of headers on the main board. The header layout is standard among the majority of the Uno-type boards and many of the basic Arduino add-ons, also known as *shields*, can be plugged directly into these headers and stacked one on top of the other, providing power and I/O directly to the shield without any additional wiring needed.

Many types of shields are available, including all manner of displays, Ethernet, Wi-Fi, motor driver, MP3, and a wide array of other devices. My personal favorite is the prototyping shield, which allows you to build your own interface to an even wider array of Arduino-compatible components, modules, and breakout boards. You can find GPS, real time clock, compass, text-to-speech, and lightning detection modules, for example, along with an endless list of sensors such accelerometers, pressure, humidity, proximity, motion, vibration, temperature, and many more. We'll explore some of these modules and sensors in projects presented in later chapters of this book.

History

As living proof that necessity is the mother of invention, the Arduino was created at the Interaction Design Institute Ivrea, in the northern Italian town of Ivrea. Originally designed as an inexpensive Open Source tool for students, replacing the more expensive and less powerful Parallax "Basic Stamp" development platform then used by students at the institute, the Arduino began as a thesis project in 2003 by artist and design student, Hernando Barragán, designed for a non-technical audience.

This project, known as *Wiring*, was based on a ready-to-use circuit board with an Integrated Development Environment (IDE) based on the *Processing* language created by Ben Fry and one of Barragán's thesis advisors, Casey Reas. Wiring was then adapted in 2005 by a team co-founded by another of Barragán's thesis advisors, Massimo Banzi. This team consisted of Hernando Barragán, Massimo Banzi, David Cuartielles, Dave Mellis, Gianluca Marino, and Nicholas Zambetti. Their goal was to further simplify the Wiring platform and design a simple, inexpensive Open Source prototyping platform to be used by non-technical artists, designers, and others in the creative field. Banzi's design philosophy regarding the Arduino is best outlined in his quote "Fifty years ago, to write software you needed people in white aprons who knew everything about vacuum tubes. Now, even my mom can program."

Unfortunately, at the same time, due a lack of funding the Institute was forced to close its doors. Fearing their projects would not survive or be misappropriated, the team decided to make the entire project Open Source. Released under the Open Source Creative Commons license, the Arduino became one of the first, if not the first, Open Source hardware products. Needing a name for the project, the team decided to name it Arduino after a local pub named "Bar Di Re Arduino" which itself honors the memory of Italian King Arduin.

Everything about the Arduino is Open Source. The board designs and schematic files are Open Source, meaning that anyone can create their own version of the Arduino free of charge. The Creative Commons licensing agreement allows for unrestricted personal and commercial derivatives as long as the developer gives credit to Arduino, and releases their work under the same license. Only the name Arduino is trademarked, which is why the various Arduino-compatible boards have names like Iduino, Ardweeny, Boarduino, Freeduino, and so on. Typically these boards are fully compatible with their official Arduino counterpart, and they may include additional features not on the original Arduino board.

Massimo Banzi's statement about the Arduino project, "You don't need anyone's permission to make something great," and Arduino team member David Cuartielles's quote, "The philosophy behind Arduino is that if you want to learn electronics, you should be able to learn as you go from day one, instead of starting by learning algebra" sums up what has made the Arduino so popular among hobbyists and builders. The collective knowledgebase of Arduino sketches and program libraries is immense and constantly growing, allowing the average hobbyist to quickly and easily develop complex projects that once took mountains of datasheets and components to build. The Arduino phenomenon has sparked the establishment of a number of suppliers for add-on boards, modules, and sensors adapted for the Arduino. The current (as of mid-2014) Arduino team consisting of Massimo Banzi, David Cuartielles, Tom Igoe, Gianluca Martino, and David Mellis has continued to expand and promote the Arduino family of products.

Since its inception, the Arduino product line has been expanded to include more powerful and faster platforms, such as the 86 MHz 32-bit Arduino Due, based on the Atmel SAM3X8E ARM Cortex-M3 processor, and the dual-processor Arduino Yun, which contains the Atheros AR9331 running an onboard Linux distribution in addition to the ATmega32u4 processor that provides Arduino functionality. With the Arduino Tre, a 1-GHz Sitara AM335x processor based Linux/Arduino dual-processor design, the Arduino now has the power needed to support processing-intensive applications and high speed communications. The Arduino variants are discussed in more detail in Chapter 2.

What is Open Source?

Generally speaking, Open Source refers to software in which the source code is freely available to the general public for use and/or modification. Probably the best example of Open Source software is the Linux operating system created by Linus Torvalds. Linux has evolved into a very powerful operating system, and the vast majority of applications that run on Linux are Open Source. A large percentage of the web servers on the Internet are Linux-based, running the Open Source *Apache Web Server*. The popular Firefox web browser is also Open Source, and the list goes on. Even the Android phone operating system is based on Linux and is itself Open Source. This ability to modify and adapt existing software is one of the cornerstones of the Open Source movement, and is what had led to its popularity and success.

The Arduino team took the concept of Open Source to a whole new level. Everything about the Arduino — hardware and software — is released under the Creative Commons Open Source License. This means that not only is the Integrated Development Environment (IDE) software for the Arduino Open Source, the Arduino hardware itself is also Open Source. All of the board design file and schematics are Open Source, meaning that anyone can use these files to create their own Arduino board. In fact, everything on the Arduino website, **www.arduino.cc**, is released as Open Source.

As the Arduino developer community grows, so does the number of Open Source applications and add-on products, also released as Open Source. While it may be easier to buy an Arduino board, shield or module, in the vast majority of cases, everything you need to etch and build your own board is freely available for you to do as you wish. The only real restriction is that you have to give your work back to the Open Source community under the same Open Source licensing. What more could a hobbyist ask for? Everything about the Arduino is either free or low cost. You have a whole community of developers at your back, creating code and projects that you can use in your own projects, saving you weeks and months of development. As you will see in some of the projects in this book, it takes longer to wire and solder things together than it does to actually get it working. That is the true power of Open Source, everyone working together as a collective, freely sharing their work, so that others can join in on the fun.

Open Source Licensing and How it Works

There are several main variations on the Open Source licensing model, but all are intended to allow the general public to freely use, modify, and distribute their work. The most common Open Source license models you will encounter include the GNU General Public License (GPL), Lesser GPL (LGPL), MIT, and the Creative Commons licenses. As a general rule, for the average hobbyist, this means you are free to do as you wish. However, there will always be those of us that come up with that really cool project we can package up and sell to finance our next idea. It is important for that group to review and understand the various license models you may encounter in the Open Source world.

The GNU GPL

As with all Open Source licensing models, the GNU General Public License (GPL) is intended to guarantee your freedom to share, modify, and distribute software freely. Developers who release software under the GPL desire their work to be free and remain free, no matter who changes or distributes the program. The GPL allows you to distribute and publish the software as long as you provide the copyright notice, disclaimer of warranty, and keep intact all notices that refer to the license. Any modified files must carry prominent notices stating that you changed the files and the date of any changes.

Any work that you distribute and publish must be licensed as a whole under the same license. You must also accompany the software with either a machine-readable copy of the source code or a written offer to provide a complete machine readable copy of the software. Recipients of your software will automatically be granted the same license to copy, distribute, and modify the software. One major restriction to the GPL is that it does not permit incorporating GPL software into proprietary programs.

The copyright usage in the GPL is commonly referred to as "copyleft," meaning that rather than using the copyright process to restrict users as with proprietary software, the GPL copyright is used to ensure that every user has the same freedoms as the creator of the software.

There are two major versions of the GPL, Version 2, and the more recent Version 3. There are no radical differences between the two versions; the changes are primarily to make the license easier for everyone to use and understand. Version 3 also addresses laws that prohibit bypassing Digital Rights Management (DRM). This is primarily for codecs and other software that deals with DRM content. Additional changes were made to protect your right to "tinker" and prevent hardware restrictions that don't allow modified GPL programs to run. In an effort to prevent this form of restriction, also known as Tivoization, Version 3 of the GPL has language that specifically prevents such restriction and restores your right to make changes to the software that works on the hardware it was originally intended to run on. Finally, Version 3 of the GPL also includes stronger protections against patent threats.

The Lesser GNU General Public License (LGPL)

The LGPL is very similar to the GPL, except that it permits the usage of program libraries in proprietary programs. This form of licensing is generally to encourage more widespread usage of a program library in an effort for the library to become a de-facto standard, or as a substitute for a proprietary library. As with the GPL, you must make your library modifications available under the same licensing model, but you do not have to release your proprietary code. In most cases, it is preferable to use the standard GPL licensing model.

The MIT License

Originating at the Massachusetts Institute of Technology, the MIT license is a permissive free software license. This license permits reuse of the software within proprietary software, provided all copies of the software include the MIT license terms. The proprietary software will retain its proprietary nature even though it incorporates software licensed under the MIT license. This license is considered to be GPL-compatible, meaning that the GPL permits combination and redistribution with software that uses the MIT License. The MIT license also states more explicitly the rights granted to the end user, including the right to use, copy, modify, merge, publish, distribute, sublicense, and/or sell the software.

The Creative Commons License

There are multiple versions of the Creative Commons License, each with different terms and conditions:

•Attribution (CC BY) — This license allows others to distribute, remix, tweak, and build upon a work, even commercially, as long as they credit the creator for the original creation.

•Attribution-NonCommercial (CC BY-NC) — This license allows others to remix, tweak, and build upon a work non-commercially. While any new works must also acknowledge the creator and be non-commercial, any derivative works are not required to be licensed on the same terms.

•Attribution-ShareAlike (CC BY-SA) — This is the most common form of the Creative Commons License. As with the Attribution license, it allows others to distribute, remix, tweak, and build upon a work, even commercially, as long as they credit the creator for the original creation, and license their new creation under the same license terms. All new works based on yours convey the same license, so any derivatives will also allow commercial use.

•Attribution-NonCommercial-ShareAlike (CC BY-NC-SA) — This license allows others to distribute, remix, tweak, and build upon a work non-commercially, as long as they credit the creator and license their new creations under the identical licensing terms.

•Attribution-No Derivs (CC BY-ND) — This license allows for redistribution, both commercial and non-commercial, as long as it is passed along unchanged and in its entirety, with credit given to the original creator.

• Attribution-NonCommercial-NoDerivs (CC BY-NC-ND) — This is the most restrictive of the Creative Commons licenses, only allowing others to download a work and share them with others as long as they credit the creator. Works released under this license cannot be modified in any way, nor can they be used commercially.

The Arduino is released under the Creative Commons Attribution-ShareAlike (CC BY-SA) license. You can freely use the original design files and content from the Arduino website, **www.arduino.cc**, both commercially and non-commercially, as long as credit is given to Arduino and any derivative work is released under the same licensing. So, if by chance you do create something that you would like to sell, you are free to do so, as long as you give the appropriate credit to Arduino and follow the requirements outlined in the FAQ on the Arduino website, as you may not be required to release your source code if you follow specific guidelines. If you include libraries in your work, be sure you use them within their licensing guidelines. The core Arduino libraries are released under the LGPL and the Java-based IDE is released under the GPL.

In Conclusion

It is this Open Source licensing that has made the Arduino so popular among hobbyists. You have the freedom to do just about anything you want and there are many others developing code and libraries you can freely incorporate in your code, which helps make developing on the Arduino so much fun. For example, I know very little about Fast Fourier transforms, but there is a fully functional library out there just waiting for me to come up with a way to use it. That's the beauty of the Arduino and Open Source. You don't have to be a programming genius to create fully functional projects as long as you have the entire Open Source developer community to draw upon. And, when you do start creating wonderful new things, remember to share them back to the community, so that others following in your footsteps can benefit from your work and create wonderful new things of their own.

References

Arduino — **www.arduino.cc**
Arduino Shield List — **www.shieldlist.org**
Atheros Communications — **www.atheros.com**
Atmel Corp — **www.atmel.com**
Creative Commons — **http://creativecommons.org**
GNU Operating System — **www.gnu.org**
Open Source Initiative — **http://opensource.org**
Texas Instruments — **www.ti.com**

CHAPTER 2

Arduino Boards and Variants

Trying to decide which of the various Arduino boards to use in your project can be confusing. Many versions of the Arduino have been developed since its creation in 2005, and there is no straightforward way to identify the capabilities and specifications of each board. In this chapter, we will briefly explore the various Arduino boards to help you determine which board best suits your project ideas.

The Arduino in General

The basic Arduino uses an Atmel ATmega-series microcontroller providing 14 digital I/O pins and six 10-bit analog-to-digital input pins that can also be used for digital I/O. Six of the digital I/O pins support pulse width modulation (PWM). Two of the digital I/O pins can be configured to support external interrupts for hardware program control and all 24 I/O pins can be configured to provide a program interrupt when the pin changes state.

The Arduino typically has three types of memory: flash, static random access memory (SRAM), and electrically-erasable programmable read-only memory (EEPROM). The flash memory is rewritable memory used to hold your Arduino programs, known as *sketches*. This memory has a lifetime of

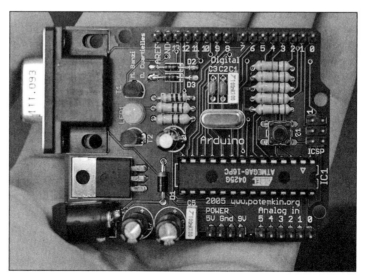

Figure 2.1 — Arduino Serial. [courtesy www.arduino.cc; Nicholas Zambetti photo]

approximately 100,000 write cycles. Flash memory can also be used to hold data that doesn't change, such as lookup tables and other constants to save valuable SRAM space through the use of the PROGMEM keyword. SRAM is used to hold your program and system variables. The EEPROM can be used to retain data between reboot or power cycles, such as calibration values and other similar data. The Arduino EEPROM also has a lifetime of approximately 100,000 write cycles.

The ATmega microcontroller supports the serial, SPI, and I^{2}C communication protocols. The majority of the Arduino boards in this series have a standard 2.7 × 2.1 inch footprint. Female headers on the edges of the board allow for the stacking of add-on interfaces, also known as *shields*, without the need for additional wiring.

The ATmega8 Series

The early Arduino boards such as the Arduino Serial shown in **Figure 2.1** were based on the 8-MHz ATmega8 processor and had 8 KB of program (flash) memory, 1 KB of static RAM (SRAM), and 512 bytes of EEPROM onboard. These early Arduino boards included the first board to bear the Arduino name, the Arduino USB, the Arduino Extreme, and the Arduino NG. Both this family and the ATmega168-based boards have been generally superseded by the ATmega328 boards, but a number of these older boards are still available.

The ATmega168 Series

The second generation of the Arduino boards used the 16-MHz Atmel ATmega168 processor. The ATmega168 increased the program memory to 16 KB, but was otherwise similar to the ATmega8. Arduino boards in this family include the first Arduino board to ship with the ATmega168, the NG+, the Arduino Bluetooth, the Nano, the LilyPad (designed to be used in wearable Arduino projects), the Mini, the Diecimila (**Figure 2.2**), and the Duemilanove.

Figure 2.2 — Arduino Diecimila. [courtesy commons.wikimedia.org; Remko van Dokkum photo]

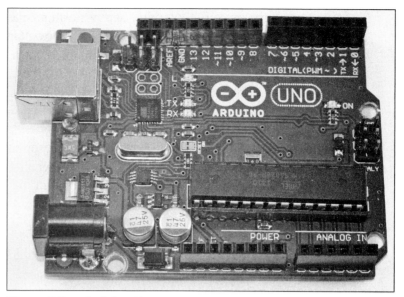

Figure 2.3 — Arduino Uno.

The ATmega328 Series

With the introduction of the 16-MHz Atmel ATmega328 processor, the Arduino's popularity began to soar, and it is this generation and its successors that this book will emphasize. The Arduino Uno R3 (**Figure 2.3**) has become the *defacto* standard for Arduino boards. Other boards in this generation include the upgraded Duemilanove, LilyPad and Nano; the Fio, which was designed for battery powered wireless applications, and the Pro-Mini. The ATmega328 has 32 KB of flash memory, 2 KB of SRAM, and 1 KB of EEPROM. In later versions, the ATmega328 processor is replaced with the ATmega32u4, identical in functionality to the ATmega328 with the addition of a USB controller integrated into the processor itself.

Bite Size Arduinos

The Arduino is available in a number of smaller footprints as well, allowing you to miniaturize your projects. These tiny Arduinos, shown in **Figures 2.4 to 2.7**,

Figure 2.4 — Arduino Pro Mini.

feature the same functionality and performance as their full-sized brothers in a much smaller form factor. The Arduino Pro Mini is a mere 0.7 × 1.3 inches and the Arduino Nano is 1.7 × 0.73 inches. Some, such as the Ardweeny from Solarbotics and the DC Boarduino from Adafruit, require the use of a USB-to-serial interface card (also known as an FTDI interface — see **Figure 2.8**) for program loading and interfacing with the IDE.

The LilyPad Arduino (**Figure 2.9**) was designed and developed by Leah Buechley and SparkFun Electronics to be used in wearable projects. A mere 2 inches in diameter, the LilyPad Arduino can be sewn into fabric with the I/O and power connections connected via conductive thread. A more recent version,

Figure 2.5 — Iduino Nano.

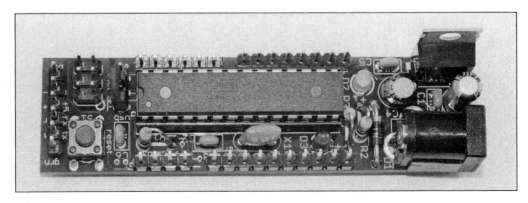

Figure 2.6 — DC Boarduino.

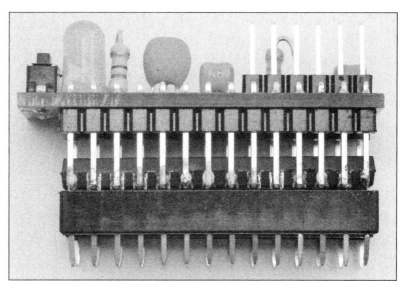

Figure 2.7 — Solarbotics Ardweeny.

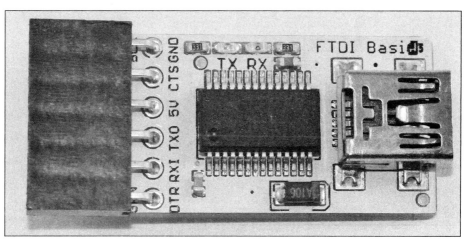

Figure 2.8 — FTDI USB to serial adapter.

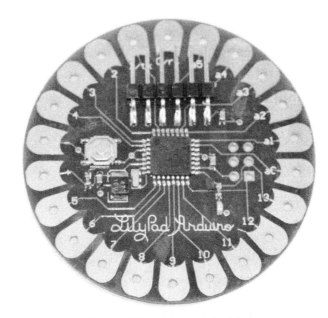

Figure 2.9 — Lilypad Arduino.

Arduino Boards and Variants 2-5

the LilyPad Arduino SimpleSnap, has a built-in rechargeable lithium polymer battery and conductive snap connectors for I/O, allowing the board to be easily removed when washing the fabric.

The Mega Series

For those needing more memory and I/O, there is the Arduino Mega series. Beginning with the Arduino Mega, based on the 16-MHz Atmel ATmega 1280, the Mega series packs a punch with 54 digital input/output pins, sixteen 10-bit analog inputs, and four serial hardware ports on a 4 × 2.1 inch board. Six of the digital I/O pins can be configured as interrupt pins and 14 digital I/O pins can provide pulse width (PWM) output. The I/O headers are arranged to be

Figure 2.10 — Arduino Mega 2560.

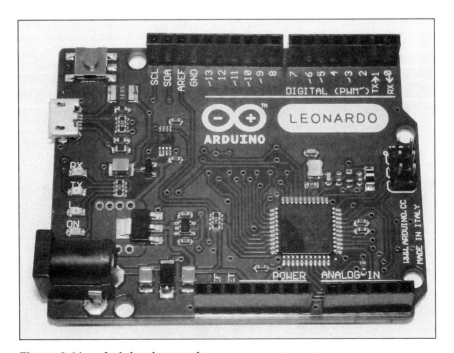

Figure 2.11 — Arduino Leonardo.

compatible with most of the Arduino shields designed for the Uno and similar Arduinos. The Arduino Mega ups the program flash memory to 128 KB, the SRAM to 8 KB, and the EEPROM to 4 KB.

The Mega was superseded by the Mega2560 and Mega2560 ADK. The 16-MHz Mega2560 (**Figure 2.10**) is based on the Atmel ATmega 2560 processor, with 256 KB of program flash memory. The Mega2560 also adds an additional digital I/O pin that can provide PWM, for a total of 15 PWM outputs. The Mega2560 ADK adds a USB host port to enable communication with Android phones for use with Google's Android Accessory Development Kit.

The New Generation of Arduinos

The Arduino continues to evolve, with newer and more powerful versions appearing on a regular basis. The latest board in the standard Arduino footprint (as of early 2014) is the Arduino Leonardo shown in **Figure 2.11**. Based on the 16-MHz Atmel ATmega32u4 processor, the Leonardo has 20 digital I/O pins. Seven of these pins can provide PWM and 12 can be 10-bit analog-to-digital pins. Five of the digital I/O pins can be configured to support interrupts. The Leonardo also has a built-in USB controller, allowing the Leonardo to appear to a connected computer as a mouse and keyboard, in addition to the standard serial COM port. The Leonardo also has 2.5 KB of SRAM, compared to the Uno's 2 KB.

The Arduino Esplora

Derived from the Arduino Leonardo, the Arduino Esplora (**Figure 2.12**) offers the same functionality and performance of the Arduino Leonardo while including a number of built-in, ready-to-use, onboard sensors. Shaped like a videogame controller, the Esplora is designed for people who want to get up and running with Arduino without having to learn about the electronics first.

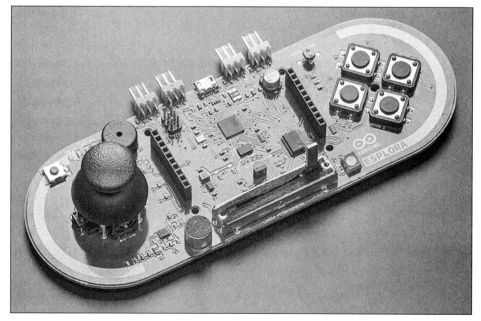

Figure 2.12 — Arduino Esplora. [courtesy Adafruit Industries (product code: 1348)]

The Esplora has a two-axis joystick with a center pushbutton, four pushbutton inputs, a slide potentiometer, microphone, light sensor, a temperature sensor, a three-axis accelerometer, a buzzer, and an RGB LED onboard. It also has with two Tinkerkit input and output connectors, along with a socket for a color TFT LCD screen. There is also a software library designed for the Esplora and its sensors, which greatly simplifies the process of interfacing with all of the onboard features.

The Arduino Due

Based on the Atmel 84-MHz SAM3X8E ARM Cortex-M3 processor, the Arduino Due shown in **Figure 2.13** is the first 32-bit Arduino microcontroller. The Arduino Due has 512 KB of flash memory and 96 KB of SRAM. Unlike previous Arduino designs, the Arduino Due does not have any onboard EEPROM. The Arduino Due has 54 digital input/output pins, twelve 12-bit analog inputs, two 12-bit digital-to-analog outputs, two I²C ports, and four hardware serial ports. Twelve of the digital I/O pins can be configured as PWM outputs. An ERASE button has been added to manually erase the contents of the flash memory. As with the Leonardo, the Arduino Due has a USB host port, which allows it to emulate a keyboard or mouse to a connected workstation.

It is very important to note that the Arduino Due runs at 3.3 V instead of the typical 5 V. The maximum voltage the I/O pins can handle is also 3.3 V. Like the Mega series, the Arduino Due is compatible with most standard Arduino shields, but care must be taken not to exceed the 3.3 V limit on the I/O pins.

Figure 2.13 —Arduino Due.

The Arduino Yún

Based on the ATmega32u4 and the Atheros AR9331, the Arduino Yún (**Figure 2.14**) is the first dual processor Arduino microcontroller board. The 16-MHz ATmega 32u4 provides Arduino functionality similar to the Arduino

Figure 2.14 — Arduino Yun. [courtesy www.arduino.cc]

Leonardo, while the 400-MHz Atheros processor supports a Linux distribution based on OpenWRT named Linino. The Arduino Yún has built-in 10/100 Mb/s Ethernet and IEEE 802.11b/g/n WiFi support, a USB, microSD card slot, 20 digital I/O pins, and twelve 10-bit analog input pins. Seven of the digital I/O pins can be configured to use pulse width modulation (PWM).

The Arduino Yún is supported by the `Bridge` library which facilitates communication between the two processors, giving Arduino sketches the ability to run shell scripts, communicate with network interfaces, and receive information from the AR9331 processor.

The Arduino Yún's 16-MHz ATmega32u4 controller has 32 KB of flash memory, 2.5 MB of SRAM, and 1 KB of EEPROM. The 400-MHz Atheros AR9331 processor on the Arduino Yún comes with 64 MB of DDR2 RAM and 16 MB of flash memory.

The Arduino Tre

Due out as this book was prepared in the spring of 2014, the Arduino Tre (**Figure 2.15**) is the first Arduino board manufactured in the US. Boasting a 1-GHz Sitara AM335x processor, the Arduino Tre offers the processing horsepower of a desktop paired with the standard Arduino functionality. Partially the result of a close collaboration between Arduino and the BeagleBoard.org foundation, the Tre design combines the benefits of both boards onto a single development platform.

The Arduino side of the Tre will offer the same features and performance of the 16-MHz Arduino Leonardo alongside the 1-GHz Texas Instrument Sitara AM3359AZCZ100 (ARM Cortex-A8) processor. The preliminary specifications

Figure 2.15 — Arduino Tre. [courtesy www.arduino.cc]

on the Sitara processor include 512 MB of DDR3 RAM, a 10/100 Ethernet port, four USB 2.0 host ports, along with HDMI video and audio ports.

The Arduino Compatibles

The success of the Arduino has led to several derivative products that warrant inclusion in this discussion about the various Arduino boards you may encounter. Digilent's chipKIT series (**Figure 2.16**) is a unique Arduino variation. Based on Microchip's 80-MHz PIC32MX320F128 processor, the chipKIT Uno32 is a 32-bit PIC microcontroller that emulates the Arduino Uno. The Uno32 has 128 KB of flash program memory and 16 KB of SRAM data memory, but has no EEPROM and is programmed using a custom version of the Integrated Development Environment (IDE) known as the Multi-Platform Integrated Development Environment (MPIDE). The MPIDE is backward-compatible with the standard Arduino IDE.

Because it emulates the Arduino Uno, the Uno32 is code compatible with most Arduino sketches, has the same basic footprint as a standard Arduino Uno, and supports most Arduino shields and devices. The chipKIT Uno32 features 30 digital I/O pins and twelve 10-bit analog-to-digital input pins. Five of the digital I/O pins can be configured to support pulse width modulation, and five digital I/O pins can be configured to support hardware interrupts. The 12 analog pins can also be configured as digital I/O pins. While the Uno32 is pin-compatible with Arduino Uno shields, the additional I/O pins on the Uno32 are brought out to a second row of header pins that are not on a standard Arduino. The chipKIT Uno32 also features two hardware serial ports, along with an onboard real-time clock calendar (RTCC). To enable the RTCC functions, a 32.768 kHz crystal must be added to the chipKIT board in the appropriate location.

The chipKIT Uno32 is a 3.3 V board, but all pins are 5 V tolerant, meaning that if you accidentally hook them to a 5 V input, no damage will occur to

Figure 2.16 — chipKIT Uno32.

the chipKIT board. Unlike the Arduino Uno, to use pins A4 and A5 for I²C communications, jumpers must be set on the chipKIT board. The chipKIT Uno32 also has a jumper-selectable option to allow the Uno32 to function as either an SPI Master or SPI Slave device.

The Digilent chipKIT Max32

The chipKIT Max32 shown in **Figure 2.17** is Digilent's variation on the Arduino Mega. Based on the 80-MHz 32-bit Microchip PIC32MX795F512, the

Figure 2.17 — chipKIT Max32. [courtesy Digilent, Inc]

Arduino Boards and Variants   2-11

Table 2.1
Arduino Comparison Chart

Board Name	CPU Type	CPU Speed	Flash Memory (KB)	SRAM (KB)	EEPROM (KB)	Digital I/O pins	Digital PWM pins	Analog Input Pins	UART	Voltage (V)	Board Footprint (inches)	Year Released
ArduinoBT	ATmega168/328	16 MHz	16/32	1/2	0.5/1	14	4	6	–	5	2.7 × 2.1	2007
Arduino Diecimila	ATmega168	16 MHz	16	1	0.5	14	6	6	–	5	2.7 × 2.1	2007
Arduino Due	AT91SAM3X8E (ARM Cortex-M3)	84 MHz	512	96	–	54	12	12	4	3.3	4 × 2.1	2012
Arduino Duemilanove	ATmega168/328	16 MHz	16/32	1/2	0.5/1	14	6	6	–	5	2.7 × 2.1	2008
Arduino Esplora	ATmega32u4	16 MHz	32	2.5	1	14	6	12	1	5	6.5 × 2.4	2012
Arduino Ethernet	ATmega328	16 MHz	32	2	1	14	4	6	–	5	2.7 × 2.1	2011
Arduino Extreme	ATmega8	16 MHz	8	1	0.5	14	4	6	–	5	2.7 × 2.1	2006
Arduino Fio	ATmega328	8 MHz	32	2	1	14	6	8	–	3.3	2.6 × 1.1	2010
Arduino Leonardo	ATmega32u4	16 MHz	32	2.5	1	14	6	12	1	5	2.7 × 2.1	2012
Arduino LilyPad	ATmega168/328	8 MHz	16	1	0.5	14	6	6	–	2.7–5.5	2 in. dia.	2007
Arduino LilyPad SimpleSnap	ATmega328	8 MHz	32	2	1	9	5	4	–	2.7–5.5	2 in. dia.	2012
Arduino Mega	ATmega1280	16 MHz	128	8	4	54	14	16	4	5	4 × 2.1	2009
Arduino Mega ADK	ATmega2560	16 MHz	256	8	4	54	14	16	4	5	4 × 2.1	2011
Arduino Mega2560	ATmega2560	16 MHz	256	8	4	54	14	16	4	5	4 × 2.1	2010
Arduino Micro	ATmega32u4	16 MHz	32	2.5	1	14	7	12	1	5	0.7 × 1.9	2012
Arduino Mini	ATmega168	8/16 MHz	16	1	0.5	14	6	6	–	3.3 or 5	0.7 × 1.3	2008
Arduino Nano	ATmega168	16 MHz	16	1	0.5	14	6	8	–	5	1.7 × 0.73	2008
Arduino Nano (version 3.0+)	ATmega328	16 MHz	32	2	1	14	6	8	–	5	1.7 × 0.73	2009
Arduino NG	ATmega8	16 MHz	8	1	0.5	14	4	6	–	5	2.7 × 2.1	2006
Arduino NG+	ATmega168	16 MHz	16	1	0.5	14	4	6	–	5	2.7 × 2.1	2006
Arduino Pro	ATmega328	8/16 MHz	32	2	1	14	6	6	–	3.3 or 5	0.7 × 1.3	2008
Arduino Pro Mini	ATmega328	8/16 MHz	32	2	1	14	6	6	–	3.3 or 5	0.7 × 1.3	2008
Arduino Tre	Sitara AM335x/ ATmega32u4	1 GHz/ 16 MHz	32	2.5	1	14	7	12	1	5	not specified	2014
Arduino UNO	ATmega328	16 MHz	32	2	1	14	6	6	–	5	2.7 × 2.1	2010
Arduino USB	ATmega8	16 MHz	8	1	0.5	14	4	6	–	5	2.7 × 2.1	2005
Arduino Yún	Atheros AR9331/ ATmega32u4	400/16 MHz	32	2.5	1	14	7	12	1	5	2.7 × 2.1	2013
Serial Arduino	ATmega 8	16 MHz	8	1	0.5	14	4	6	–	5	2.7 × 2.1	2005
Adafruit DC Boarduino	ATmega328	16 MHz	32	2	1	14	6	6	–	5	3 × 0.8	2007
Adafruit USB Boarduino	ATmega328	16 MHz	32	2	1	14	6	6	–	5	3 × 0.8	2007
Solarbotics Ardweeny	ATmega1280	16 MHz	32	2	1	14	6	6	–	5	1.6 × 0.54	2010
Microchip chipKIT Max32	Microchip PIC32MX320F128	80 MHz	128	16	–	30	5	12	2	3.3	2.7 × 2.1	2011
Microchip chipKIT Uno32	Microchip PIC32MX795F512	80 MHz	512	128	–	67	16	5	4	3.3	4 × 2.1	2011

Max32 has the same form factor as the Arduino Mega series and is compatible with many Arduino shields, as well as the larger shields designed for the Arduino Mega boards. The Max32 has 512 KB of flash program memory and 128 KB of SRAM data memory, but has no onboard EEPROM. The Max32 has 67 digital I/O pins, sixteen 10-bit analog-to-digital I/O pins than can also be configured to be digital I/O pins, and four hardware serial ports.

As with the chipKIT Uno32, the Max32 is a 3.3 V board. All pins are 5 V tolerant. The chipKIT Max32 also has a jumper-selectable option to allow the board to function as either an SPI Master or SPI Slave device. As with the Uno32, five of the digital I/O pins can be configured to support pulse width modulation and five digital I/O pins can be configured to support hardware interrupts for hardware program control. The Max32 also has an onboard real-time clock calendar (RTCC). To enable the RTCC functions, a 32.768 kHz crystal must be added to the board in the appropriate location.

The Max32 includes several features not found on the chipKIT Uno32. These include a USB On-The-Go (OTG) controller that allows the Max32 USB port to act as a USB device, USB host, or USB OTG host/device. With the addition of transceiver components, the Max 32 provides a 10/ 100 MB/s Ethernet port and two controller area network (CAN) ports. CAN is a networking standard that was originally developed for use in the automotive industry and is now finding its way into building automation and other industrial applications.

The Max32 is programmed using the same custom version of the Integrated Development Environment (IDE) used with the Uno32, the Multi-Platform Integrated Development Environment (MPIDE). The MPIDE is backward-compatible with the standard Arduino IDE allowing most sketches written for the Arduino to run on the Max32.

With so many versions of the Arduino and Arduino-compatible boards, the comparison chart in **Table 2.1** can be used to help you decide which board best suits your project's needs

References

Adafruit Industries — **www.adafruit.com**
Arduino — **www.arduino.cc**
Atmel Corp — **www.atmel.com**
Atheros Communications — **www.atheros.com**
Diligent — **www.digilentinc.com**
Solarbotics — **solarbotics.com**
SparkFun Electronics — **www.sparkfun.com**
Texas Instruments — **www.ti.com**

CHAPTER 3

Arduino Shields, Modules, and Devices

The power and simplicity of the Arduino is its ability to interface to a wide variety of sensors and devices. While the list of devices you can interface to an Arduino is seemingly endless and growing day by day, assembled here is a list of the various Arduino shields, modules, and devices I have found to be useful in ham radio projects.

Shields

As you begin to create projects with the Arduino, you may want to start out with preassembled shields rather than design and wire up individual modules and devices. Shields are boards designed to plug into the headers on the Arduino main board and allow instant access to the features on the shield without having to do any additional wiring. Multiple shields can be stacked one on top of the other, adding functionality with each stacked shield; however, the result can become cumbersome rather quickly and there is the risk of pin usage conflicts between the various shields. Typically, you'll want to have only one or

Figure 3.1 — 16 character by 2 line (16×2) LCD display. [courtesy Adafruit Industries (product code: 772)]

two shields stacked on the Arduino board. When your project expands beyond that, it is best to use a protoshield (prototyping shield) or some other method to move your project off the unwieldy shield stack and onto something more functional. Shields are generally supported by program libraries and examples, which allow you to quickly and easily develop working sketches.

LCD Display Shield

Probably the first shield you will use is an LCD display shield (**Figure 3.1**). These shields integrate a Hitachi HD44780-compatible 16-character by 2-line (16×2) LCD display onto the shield board. There are two main derivations on the LCD shield. One shield variation communicates with the LCD using seven digital I/O lines, while the other uses the I²C bus to communicate with the LCD serially using the SDA (pin A4) and SCL (pin A5) pins. Both versions have four user-programmable pushbutton switches to provide input to the Arduino.

The Graphic LCD4884 Shield

The DFRobot Graphic LCD4884 shield (**Figure 3.2**) is an 84×48 pixel graphic LCD display. Also known as the Nokia 5110 display, it communicates via the shield interface using the SPI bus and can display both text and graphics. The DFRobot shield also integrates a miniature five-way joystick on the shield board along with six digital I/O and five analog I/O pins brought out to headers on the shield itself.

Figure 3.2 — Nokia 5110 LCD shield. [courtesy. DFRobot (product code: DFR0092)]

Figure 3.3 — Color TFT shield. [courtesy Adafruit Industries (product code: 802)]

Color TFT Display Shield

New on the scene for the Arduino are the color graphic thin-film-transistor (TFT) displays such as the one shown in **Figure 3.3**. These 18-bit (262,144 shade) color displays are currently available in 1.8 inch (128×160 pixels) and 2.8 inch (240×320 pixels). The display communicates with the Arduino via the SPI bus and includes an onboard SD card slot you can use to hold and display images. The 1.8 inch version of the shield also includes a five-way joystick, while the 2.8 inch version incorporates a resistive touch screen. Using their powerful libraries, these displays allow you to rotate the entire screen and draw pixels, lines, rectangles, circles, rounded rectangles, and triangles. Of course they also display normal text.

The 4D Systems adapter shield allows you to use a variety of color TFT displays from 2.4 inches all the way up to 4.3 inches. The adapter shield interfaces via the Arduino serial interface and uses serial commands to display images, draw lines, rectangles, circles, and show text. Using the Arduino libraries and the 4D Systems Workshop Integrated Development Environment (IDE), you can create 4D Graphics Language (4DGL) display applications quickly and easily.

Relay Shield

Controlling things is one of the reasons we got the Arduino in the first place. Relays allow you the freedom to control high voltage and high current devices that otherwise would turn your Arduino into a pile of smoldering mush if you tried to connect them directly. The DFRobot relay shield (**Figure 3.4**) allows you to control four onboard relays and includes test buttons and indicator LEDs.

Figure 3.4 — Relay shield. [courtesy DFRobot (product code: DFR0144)]

Figure 3.5 — Motor shield. [courtesy DFRobot (product code: DRI0001)]

As with the relay shield, the motor driver shield shown in **Figure 3.5** allows you to control motors from your Arduino. The Arduino motor driver shield can drive two bidirectional dc motors or one bipolar stepper motor.

Audio Shields

At some point you'll want your Arduino to make sounds or play audio in response to the sensor inputs. With one of the various audio shields, you can quickly and easily add high quality sound and music to your Arduino sketches.

The Adafruit wave shield (**Figure 3.6**) can play up to 22-kHz, 12-bit WAV audio files that are stored in the onboard SD memory card. The SparkFun MP3 player shield can store and play MP3 files from its onboard microSD memory card.

Ethernet Shield

The Arduino Ethernet shield (**Figure 3.7**) gives your Arduino access to the Internet. With its powerful libraries and examples, you can turn your Arduino into a web-enabled simple chat server, web server or telnet client. It can make

Figure 3.6 — Adafruit wave shield. [courtesy Adafruit Industries (product code: 94)]

Figure 3.7 — Ethernet shield.

http requests and much more. The Ethernet shield also has an onboard microSD adapter that can be used to store files and other data. Some of the Ethernet shields also include an option to install a power-over-Ethernet (POE) module, allowing your Arduino to be powered by other POE-enabled Ethernet devices.

The USB host shield (**Figure 3.8**) turns your Arduino into a USB host device, allowing you to communicate with USB devices such as keyboards, mice, joysticks, game controllers, ADK-capable Android phones and tablets, digital cameras, USB sticks, memory card readers, external hard drives, and Bluetooth dongles.

WiFi Shield

The Arduino WiFi shield (**Figure 3.9**) allows your Arduino to connect

Figure 3.8 — USB host shield.

Figure 3.9 — Arduino WiFi shield. [courtesy www.arduino.cc]

to IEEE 802.11b/g wireless networks and supports WEP and WPA wireless encryption. As with the Ethernet shield library, the WiFi library and examples allow you to turn your Arduino into a web-enabled simple chat server, web server, or telnet client. It can make http requests and much more. The WiFi shield also has an onboard microSD adapter that can be used to store files and other data.

SD Card Shields

The SD card shields such as those shown in **Figures 3.10** and **3.11** allow you to interface standard and microSD cards to the Arduino. With one of these cards you can add mass-storage capabilities for FAT-16 and FAT-32 file access and data logging. Some versions of the SD card shields also have an onboard real-time clock calendar (RTCC) allowing you to timestamp your data and provide date and time information for your Arduino sketches.

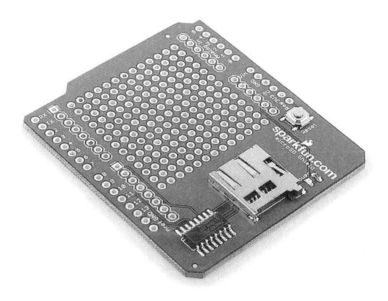

Figure 3.10 — Solarbotics microSD shield. [courtesy Solarbotics (product code: 50833)]

Figure 3.11 — Adafruit data logging shield. [courtesy Adafruit Industries (product code: 1141)]

GPS Logger Shield

The Adafruit Ultimate GPS logger shield (**Figure 3.12**) combines a GPS and a microSD card to create a full-featured GPS and data logging shield. The libraries and examples included allow parsing of the GPS standard NMEA "sentences" to read GPS data such as date, time, latitude, longitude, altitude, speed, and other standard GPS data. In addition to data logging capability, the SD library can be used to store files and other data.

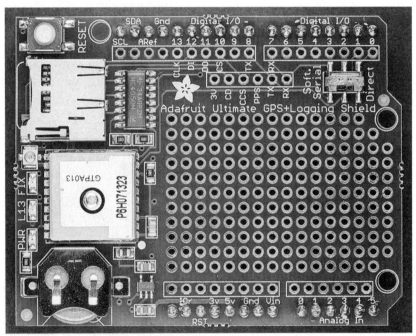

Figure 3.12 — Adafruit GPS logger shield. [courtesy Adafruit Industries (product code: 98)]

Xbee Shield

The Xbee shield (**Figure 3.13**) allows an Arduino board to communicate wirelessly using the ZigBee protocol. ZigBee is based on the IEEE 802.15 standard and is used to create a small "personal area network" suitable for communication with remote sensors or devices. The Xbee shield can communicate at a speed of 250 Kb/s at distances up to 100 feet indoors or 300 feet outdoors, and can be used as a serial/USB replacement or in a broadcast/mesh network.

Argent Data Radio Shield

The Argent Data Radio shield can be used to provide AX.25 packet radio send and receive capability to the Arduino. Packets are sent and received in AX.25 UI frames at 1200 baud allowing operation on the VHF APRS network. There are a number of projects that utilize this board in ARRL's *Ham Radio for Arduino and PICAXE* by Leigh Klotz, WA5ZNU (**www.arrl.org/shop**, ISBN: 978-0-87259-324-4). The Argent Data Radio shield also includes a prototyping area, and an HD44780-compatible LCD interface.

Figure 3.13 — Solarbotics Xbee shield. [courtesy Solarbotics (product code: 50835)]

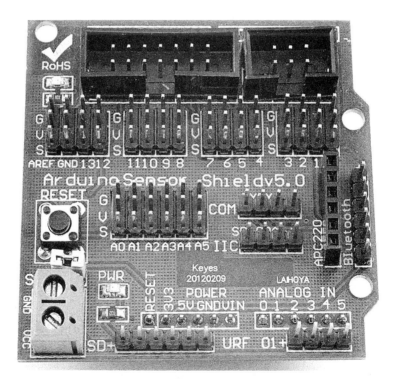

Figure 3.14 — Arduino Uno sensor shield.

I/O Shields

The I/O shield is a great way to begin creating your own projects on the Arduino (see **Figures 3.14** and **3.15**). The I/O shield brings all of the Arduino pins out to groups of 3-pin headers, along with power and ground on each header. This allows you to easily connect the Arduino I/O pins to your breadboard and prototype boards for development. In the case of the Uno I/O shield some of the pins are arranged by typical function, such as parallel LCD, serial LCD, I^{2}C, COM, and so on. Once your development is complete, all you have to do is remove the connecting wires and your Arduino is ready for the next project.

Figure 3.15 — Arduino Mego sensor shield.

Breadboard Shield

As you grow more confident and experienced in your Arduino adventures, you'll want to be able to prototype and test your designs. For the smaller projects that only involve a chip or two, I have found the breadboard shield (**Figure 3.16**) to be a quick and simple solution. Smaller than a full size

Figure 3.16 — Arduino breadboard shield.

breadboard and without the need for soldering, the breadboard shield allows quick and simple setup and redesign of your project prior to taking it full scale. When you're done, just as in a full size breadboard, simply remove the chips and wires and it's ready for the next project.

Prototyping Shield

The final step in your project will usually require a more permanent solution involving wiring and soldering components. I have found the easiest method to interface my projects to the Arduino is to use the prototyping shield (**Figure 3.17**). The protoshield brings all of the Arduino pins and voltages to solder pads with the center of the board laid out like a standard prototyping board. I like to solder header pins in the prototyping area so that I can wire the Arduino pins to the header, and then use a connector to interface to my off-board modules and components. This allows for easy troubleshooting, reconfiguration, and replacement of the external parts when I happen to let the smoke out of them occasionally.

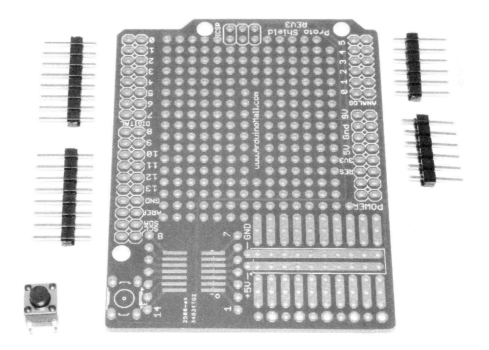

Figure 3.17 — Arduino Uno protoshield.

Arduino Modules

While shields provide a quick and easy way to interface with the Arduino, the real fun in the Arduino is interfacing with modules and additional components. It is here that the real blending of computing and electronics takes place. With just a handful of modules and parts, you can wire up some pretty amazing projects. Most of the modules include libraries, which greatly simplify communication between the Arduino and module, and you can create fully functional projects in just a matter of hours.

Using individual modules and components also lessens the pain, if and when you do horrible things to your project. Shields tend to be more expensive than individual modules and components. Since many shields use surface mount components, when something goes wrong, usually the entire shield is dead and you have to go buy another one. With the lower cost of individual modules and components, fewer tears are shed and if you're like me, you probably have more than one willing victim in your now rapidly growing supply of parts.

Fortunately, the Arduino, shields, and modules have proven to be quite forgiving when miswired. As long as you pay attention to the 3.3 and 5 V power supply differences between some of the modules and components, you have to really work to damage them. I'm not saying it can't be done, and you would be surprised at just how much smoke a teeny-tiny surface mount chip has inside, but you really have to make an effort to make it happen.

Another major advantage of modules over shields is that projects using shields tend to be bulky and unwieldy. Modules give you control over the size and shape of your project, allowing you to determine what the end product will look like. To this end, there are also a number of enclosures designed for the Arduino and Arduino projects that can really spice up the look of your completed project. In this section we will introduce and discuss the various modules that I have found to be most useful for creating ham radio projects.

Displays

There are many display modules available for the Arduino, from the easy to use basic two line text displays all the way up to the newer color graphic and ePaper displays.

16-character by 2-line LCD Display

The most commonly used display for the Arduino is the Hitachi HD44780-compatible 16-character by 2-line (16×2) LCD shown in **Figure 3.18**. A larger 16-character by 4-line (16×4) version of this display, shown in **Figure 3.19**, is also available. Both types are interfaced using six digital I/O pins on the Arduino — two pins for control, and four pins for data. Example sketches for using this display are included in the Arduino Integrated Development Environment (IDE).

A variation on the Hitachi HD44780-compatible LCD adds an I²C "backpack" module to the standard LCD, allowing you to communicate with the display using the I²C bus. This display requires only the Arduino SDA and

Figure 3.18 — 16 character by 2 line (16×2) LCD display.

Figure 3.19 —16 character by 4 line (16×4) LCD display.

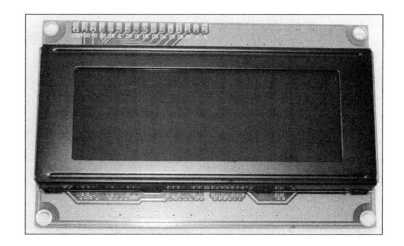

Figure 3.20 — 16 character by 2 line LCD display with I²C backpack.

Figure 3.21 — 16 character by 4 line LCD Display with I²C backpack.

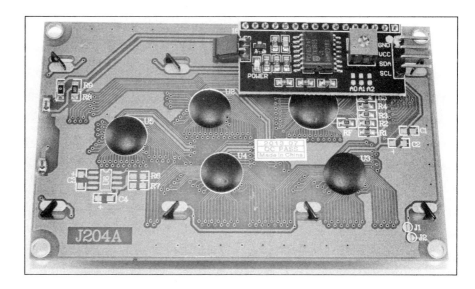

SCL pins, and can share the I²C bus with other devices. As with the standard Hitachi HD44780-compatible LCD, this display type is available in both 2 and 4-line versions (**Figures 3.20** and **3.21**).

20 × 2 Vacuum Fluorescent Display

If you want to give your Arduino display the "retro" look, you might want to use a vacuum fluorescent display (VFD). The Samsung 20T202DAJA series

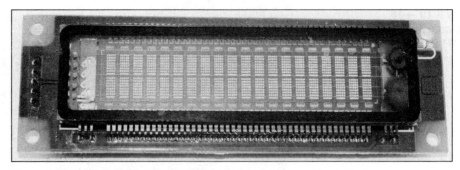

Figure 3.22 — 20 character by 2 line vacuum fluorescent display.

is a 20-character by 2-line VFD display (**Figure 3.22**). This display interfaces via SPI and is similar in display quality to the 16×2 LCDs, except that the VFD is much brighter and can be seen in sunlight. Because it is a fluorescent display, no backlight is needed.

Nokia 5110 Display

Originally used in the older Nokia cell phones, the Nokia 5110 84×48 pixel graphic LCD shown in **Figure 3.23** is becoming one of the most popular Arduino displays. Easy to interface to the Arduino using five digital I/O pins, the Nokia 5110 is small, but easily readable. With the basic library using the small font, you can display up to six lines of 14 characters each. There are many different libraries available for this display, with some supporting the 84×48 pixel graphics capability. The Nokia 5110 is based on the Phillips PCD8544 LCD controller, and while specified for 3.3 V, the Nokia 5110 will work with 2.7 to 5 V dc without damage to the display. The contrast settings can vary from display to display and most libraries allow you to control the contrast via software with simple library commands. The Nokia 5110 module also has a

Figure 3.23 — Nokia 5110 LCD display.

four LED backlight that can be controlled through use of an additional digital I/O pin.

Organic LED (OLED) Displays

The organic LED (OLED) displays are small graphic LED displays that can communicate with the Arduino using either the SPI or I²C bus. The OLED displays for the Arduino come in two versions, 128×32 pixels (**Figure 3.24**) and 128×64 pixels (**Figure 3.25**). An OLED display is essentially comprised of tiny individual LEDs, and since it uses LED technology, does not need a backlight. The libraries for the OLED displays allow you to show both text and graphics. Using the small font, two or four lines of 21 characters per line can be displayed depending on the OLED version. The OLED is an ideal choice when a small, bright, and clearly readable display is desired.

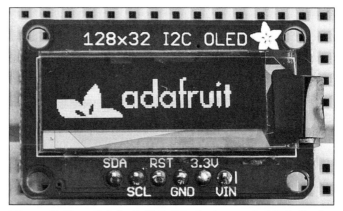

Figure 3.24 — Adafruit 128×32 pixel organic LED display. [courtesy Adafruit Industries (product code: 931)]

Figure 3.25 — Adafruit 128×64 pixel organic LED display. [courtesy Adafruit Industries (product code: 938)]

Color TFT Displays

To really spice up your Arduino display, you can use a graphic color thin-film-transistor (TFT) display. The typical color TFT display for the Arduino is available in sizes from 1.8 inches all the way to 4.3 inches.

The Adafruit color TFT displays such as the one shown in **Figure 3.26** are currently available in 1.8 (128×160 pixels), 2.2 (320×240 pixels) and 2.8 inches (240×320 pixels), and offer 18-bit color resolution, yielding 262,144 different color shades. These displays communicate with the Arduino using the SPI interface and include an onboard microSD card slot. The 2.8 inch TFT from Adafruit also has a touch screen interface, allowing you to really spice up your Arduino projects.

The intelligent color TFT displays from 4D Systems are currently available in 2.4 inches (240×320 pixels), 2.8 inches (240×320 pixels), 3.2 inches (240×320 pixels), 4.3 inches (480×272 pixels), and a whopping 7 inches (800×480 pixels), all capable of displaying more than 65,000 different colors and include built-in touch screens and microSD card slots. The 4D Systems TFT modules communicate over a standard serial port and are supported by an extensive software library. Powered by the 4D Systems PICASO processor, these modules also include additional onboard features such as two hardware serial ports, 13 additional digital I/O pins, eight 16-bit timers, an I^{2}C master interface, 14 KB of flash memory, and 14 KB of SRAM that is available for program use.

Figure 3.26 — Adafruit color TFT module. [courtesy Adafruit Industries (product code: 1480)]

4D Systems VGA Display Modules

Okay, admit it. You've fantasized about what it would be like to have your Arduino display on a standard VGA monitor. That was one of my earliest "Wouldn't it be cool if..." moments with the Arduino. Then I discovered the 4D Systems uVGA-II module (**Figure 3.27**) and that dream instantly became

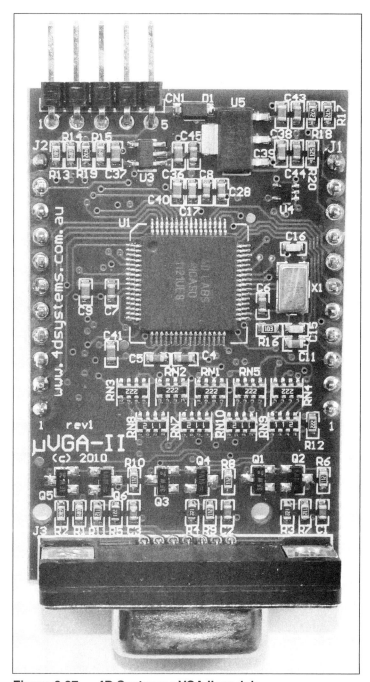
Figure 3.27 — 4D Systems uVGA II module.

a reality. Since upgraded to the uVGA-III, this standalone module accepts standard serial commands and displays the output on a standard VGA monitor. The uVGA-III supports resolutions of 320×240, 640×480, and 800×480 with 65,000 colors. Powered by the 4D Systems PICASO processor, the uVGA-III includes 15 KB of flash memory and 14 KB of SRAM onboard available for program use. The uVGA-III also includes two hardware serial ports, an I²C master interface, 13 additional digital I/O pins, eight 16-bit timers, and a microSD card slot. The uVGA-III comes with an extensive graphics library and outputs to a standard 15-pin VGA monitor interface.

Arduino Shields, Modules, and Devices   3-17

As you can see, there are a number of display options available for the Arduino, but the list of modules you can attach to your Arduino is not limited to just displays.

FTDI Module

The FTDI module (named after Future Technologies Devices International — FTDI — chip used on the module), is used to communicate with an Arduino board that does not support a USB connection, such as the Ardweeny and DC Boarduino. The FTDI module converts the USB data from your workstation to standard TTL serial signals used by the non-USB Arduino. See **Figure 3.28**.

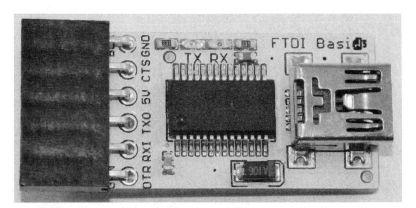

Figure 3.28 — FTDI USB-to-serial adapter.

Level Converter Module

You may run into the situation where you have a 3.3 V Arduino, such as the Arduino Due, and need to connect to a 5 V module. Or you might have a 3.3 V module that you need to connect to an Arduino Uno. If you connect these directly, you may cause damage to the Arduino or the module. Fortunately, you can use a level converter module to quickly and easily convert between the

Figure 3.29 — 3.3 V to 5 V level converter.

two signal voltage levels. The SparkFun logic level converter module shown in **Figure 3.29** can convert between 3.3 V and 5 V levels on four pins (two input and two output). It can also be used to adapt 1.8 V and 2.7 V devices to 5 V.

Weather Sensors

Many hams have an interest in monitoring weather conditions, whether for emergency preparedness or just plain old weather watching. The Arduino has a wide array of sensors that can be used to monitor various weather-related data such as temperature, relative humidity, barometric pressure, and other weather-related information.

DS18B20 Temperature Sensor

The Maxim DS18B20 (**Figure 3.30**) is a self-contained single chip temperature sensor that provides Celsius temperature readings from –55 to +125 °C (–67 to +257 °F) with an accuracy of ±0.5 °C over most of that range, and with a programmable resolution of 9 to 12 bits. The DS18B20 communicates with the Arduino using the Maxim 1-Wire interface and can be located as far away as 200 meters. The DS18B20 can either be powered normally with 3 to 5 V or powered parasitically, drawing power from the data line itself. The DS18B20 also has a programmable nonvolatile alarm capability, based on upper and lower alarm trigger points. The `OneWire` and `Dallas Temperature Control` Arduino program libraries greatly simplify communication between the Arduino and 1-Wire devices such as the DS18B20. Each 1-Wire device has a unique 64-bit serial number embedded on the chip, allowing multiple temperature sensors to be accessed using a single digital I/O pin on the Arduino. The `OneWire` library supports up to 127 1-Wire devices attached to the 1-Wire bus.

RHT03 Humidity/Temperature Sensor

The MaxDetect RHT03 shown in **Figure 3.31** (also known as the DHT-22) is a single-component humidity and temperature sensor that can provide 16-bit relative humidity (RH) readings with a resolution of 0.1% RH and an accuracy of ±2%. It also provides 16-bit temperature readings from –40 to +80 °C (–40 to +176 °F) with a resolution of 0.1 °C and an accuracy of ±0.5 °C. The RHT03 is powered with 3.3 to 6 V and communicates with the Arduino at a distance of up to 100 meters via a single digital I/O pin using the MaxDetect 1-Wire Bus protocol. It is important to note that the MaxDetect 1-Wire Bus is not compatible with the Maxim 1-Wire Bus used for devices such as the DS18B20 temperature sensor. Fortunately, there are several excellent Arduino libraries for the RHT series of sensors that add functionality to the RHT03, including the capability to calculate dew points directly.

Figure 3.30 — Dallas Semiconductor (now Maxim Integrated) DS18B20 temperature sensor.

Figure 3.31 — MaxDetect RHT03 relative humidity and temperature sensor module.

Figure 3.32 — Bosch BMP085 barometric pressure and temperature sensor module.

BMP085 Barometric Pressure Sensor

The Bosch BMP085 barometric pressure sensor module measures barometric pressure from 300 to 100 hPa (hectopascals) with an accuracy of ±1.5 hPa and a resolution of 0.01 hPa. The BMP085 also includes a temperature sensor with a range of –40 to +85°C and an accuracy of ±2 °C at a resolution of 0.1 °C. The BMP085 is self-calibrating and operates with a supply voltage of 1.8 to 3.3 V (some modules also support 5 V) and communicates with the Arduino using the I²C bus. The program libraries for the BMP085 add the capability to directly calculate altitude from the BMP085 data.

Direct Digital Synthesizer Modules

One of the more interesting modules for the Arduino is the direct digital frequency generator (DDS) module such as the AD9833 module shown in **Figure 3.33**. Using a DDS module, you can programmatically generate highly stable waveforms of varying frequencies from a single reference clock. You can use a DDS to generate various waveforms, including sine waves, square waves, and in the case of some DDS modules, even triangle waves. The major advantage of using a DDS module over processor-generated waveforms is that the DDS chip offloads all of the frequency and waveform generation from the processor and uses a 10-bit digital-to-analog converter to produce clean, stable waveforms across the entire operating range.

While there are a number of DDS modules available, the three modules I prefer to use in projects are based on the Analog Devices AD9833, AD9850, and AD9851 DDS chips. The AD9833 programmable waveform generator

Figure 3.33 — Analog Devices AD9833 waveform generator module.

has dual 28-bit frequency control registers and can generate sine, square, and triangle waves from 0 to 12.5 MHz with a resolution of 0.1 Hz. The AD9850 (**Figure 3.34**) has a single 32-bit frequency control register and can generate sine and square waves from 0 to 62.5 MHz with a resolution of 0.0291 Hz. Finally, the AD9851 (**Figure 3.35**) has a single 32-bit frequency control register and can generate sine and square waves from 0 to 70 MHz with a resolution 0.04 Hz. All of these DDS modules also include the capability of shifting the phase of the output waveform from 0-2π (0-720°) and are interfaced to the Arduino using the SPI bus. While the software libraries available for the DDS modules are somewhat lacking at the current time, there is a wide selection of sample code available to help you on your way.

Figure 3.34 — Analog Devices AD9850 direct digital synthesis (DDS) module.

Figure 3.35 — Midnight Design Solutions DDS-60 module using the AD9851 chip.

Figure 3.36 — SKYLAB SKM53 GPS module.

GPS Module

With a GPS module such as the one shown in **Figure 3.36**, you can add all of the power of a standard GPS to your Arduino, providing such information as time (down to 1/100 of a second), latitude, longitude, altitude, speed, course, and many other features of a standard GPS. These GPS modules communicate with the Arduino via a standard serial I/O port, typically with a default speed of 9600 baud and they output standard National Marine Electronics Association (NMEA) NMEA-0183 messages. The TinyGPS and TinyGPS+++ program libraries parse the NMEA messages sent by the GPS into data that can be accessed using simple function calls from within your sketch.

Emic 2 Text-to-Speech Module

Designed by Parallax in conjunction with Grand Idea Studios, the Emic 2 text-to-speech module (**Figure 3.37**) allows you to add natural sounding speech to your Arduino projects. Capable of speaking in English and Spanish, the Emic 2 features nine preprogrammed voice styles in addition to program control of speech characteristics such as pitch, speaking rate, and word emphasis.

The Emic 2 communicates with the Arduino using a standard serial I/O port at a default speed of 9600 baud and includes an onboard audio amplifier and audio jack. No program libraries are needed. All you have to do is send the text to the serial port and the Emic 2 converts it, along with any speech characteristic commands, directly into speech.

Figure 3.37 — Parallax/Grand Idea Studios Emic 2 text-to-speech module.

Arduino Shields, Modules, and Devices   3-23

Figure 3.38 — Embedded Adventures AS3935 MOD-1016 lightning detector module.

Lightning Detector Module

The lightning detector module shown in **Figure 3.38** is based on the Austriamicrosystems Franklin AS3935 lightning detector sensor and can detect lighting at a distance up to 40 km. The AS3935 is capable of detecting both cloud-to-ground and cloud-to-cloud lightning. It includes an embedded algorithm to reject man-made electrical noise and has program controllable threshold settings. The AS3935 will statistically calculate the distance to the leading edge of the storm and the estimated strength of the lightning strike. It can communicate with the Arduino via either the SPI or I²C bus.

Digital Compass Module

The HMC5883L triple axis magnetometer module (**Figure 3.39**) allows you to integrate the functions of a digital compass into your Arduino projects. Through the use of magneto-resistive sensors, the module can calculate the magnetic force on the sensors and determine magnetic North. The HMC5883L communicates with the Arduino via the I²C bus. Be careful with the supply voltage, as the basic HMC5883L is a 3.3 V device, although some modules now support 5 V.

Current Sensor Module

The INA169 current sensor module (**Figure 3.40**) allows you to measure dc current flow in a circuit. The

Figure 3.39 — HMC5883L 3-axis digital compass module.

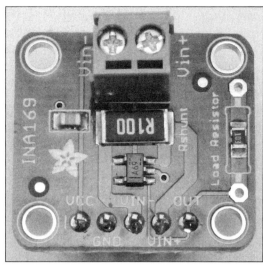

Figure 3.40 — INA169 current sensor module.

Figure 3.41 — JY-MCU Bluetooth slave module.

Texas Instruments INA169 chip converts the dc current flow across a shunt resistor into an output voltage that corresponds to 1 V/A (volt per ampere) that can be read using an analog input pin on the Arduino. Capable of monitoring up to 5 A at a maximum of 60 V continuously, the load resistor on the module can be changed to increase or decrease the sensitivity of the sensor.

Bluetooth Module

There are two forms of Bluetooth modules, slave and master. See **Figure 3.41**. A Bluetooth slave module allows you to pair and connect your Arduino to a device such as a workstation or cell phone and have the Arduino act as a mouse, keyboard, or any other human interface device (HID). A Bluetooth master module allows you to pair and connect multiple Bluetooth slave modules to your Arduino.

Some of the currently available Bluetooth modules can be switched between slave and master mode from within your Arduino sketch. The

Arduino communicates with the Bluetooth module using standard TTL serial communication, usually at a speed of 9600 baud.

Be careful and look up the specifications for your particular Bluetooth module, as some accept power in the range of 3.6 to 6 V, but only a maximum of 3.3 V on the data pins. If you apply a 5 V digital I/O pin to a 3.3 V Bluetooth module input pin, chances are you've just fried your module and it will be time to dig another one out of your parts supply (I learned this lesson the hard way). You can use a simple voltage divider network or a level converter module to protect the data pins from overvoltage.

Real-Time Clock Calendar Module

The Arduino does not have a real clock onboard. It has a timer that keeps track of the number of milliseconds since the last reset, power on, or sketch upload. You can use this timer to keep track of time, but since it is volatile, it isn't much use after a reset. To solve this problem, and to give your Arduino true clock/calendar functionality, you can use a real-time clock/calendar module (RTCC) such as the one shown in **Figure 3.42**. Many of the RTCC modules are based on the Maxim DS1307 RTCC chip and communicate with the Arduino using the I²C bus. This chip keeps track of time, day, month and year. It includes leap year compensation and a calendar accurate to the year 2100 (at which time there will be a huge market for Arduino Y2.1K bug fixers — mark your calendars).

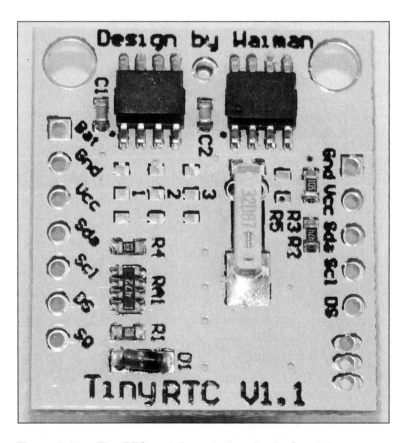

Figure 3.42 — TinyRTC real-time clock calendar/module.

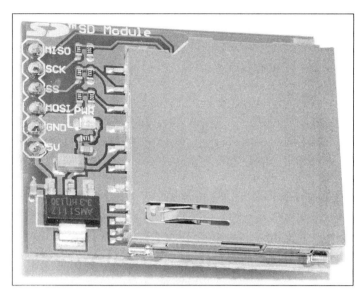
Figure 3.43 — SC card module.

Backed up by a lithium coin-cell battery, the DS1307 can run and retain data for a minimum of nine years. The DS1307 also contains 56 bytes of battery-backed-up nonvolatile RAM (NVRAM) that can be used to store data. The newer libraries for the DS1307 include support for reading and writing to the 56 bytes of NVRAM.

SD Card Modules

SD card readers allow the use of SD and microSD cards with your Arduino for read/write file access and data logging. While some of the various shields and other modules include an SD card slot, often you cannot access the SD card while using the primary device (often a TFT display). By using a standalone SD card reader (**Figure 3.43**), you can communicate with the SD card and your other SPI devices without this limitation. The standard SD and microSD card modules communicate with the Arduino via the SPI bus and include onboard level shifters to allow use at 3.3 or 5 V.

Motor Driver Module

The L298N dual motor driver module (**Figure 3.44**) has two onboard H-bridges to control up to two bidirectional dc or stepper motors. The STMicroelectronics L298N can handle up to 50 V at 2 A per motor and is controlled using six digital I/O lines (three per motor). Some modules also include current sensing capability, allowing you to sense the current draw of the motor while it is active.

Analog-to-Digital Converter Modules

The analog-to-digital (A/D) converter in the Arduino is a single-ended 10-bit A/D, providing a count of 0 to 1023 over a 5 V range with respect to

Figure 3.44 — Stepper motor driver board.

Figure 3.45 — Texas Instruments ADS1115 4-channel 16-bit analog-to-digital converter module.

ground. For many projects, this resolution is more than adequate; however, there will be occasions where you will want higher resolution on your analog measurements. In the case of an antenna rotator position sensor, you will want to read the position from 0 to 360° (or 450° in the case of some rotators). Using a 10-bit A/D, this leaves you with a resolution of about 1°. A 12-bit A/D can provide you with a resolution of about $1/10°$, and a 16-bit A/D will provide a resolution of about $1/100°$. As you can see, the more bits your A/D has, the higher the resolution. While you may not need that level of resolution, in this case, it will help to position your antennas more accurately.

The Texas Instruments ADS1015 12-bit A/D module can provide a range of 0 to 4096 on four single-ended or two differential inputs with sampling speed selectable up to 3300 samples per second. The Texas Instruments ADS1115 16-bit A/D module (**Figure 3.45**) can provide a range of 0 to 65535 on four single-ended or two differential inputs with sampling speed selectable up to 860 samples per second. Both communicate with the Arduino using the I²C bus and feature six program-selectable gain settings from $2/3$ to 16, allowing you to maximize the resolution of the attached sensors. The Arduino program libraries for these modules allow you to control all of the internal parameters of the A/D chips and make using these modules in your sketches a snap.

Digital-to-Analog Converters

One thing I wish the Arduino had is a digital-to-analog (D/A) converter output. In my lab, I have built multiple versions of D/A converters for the Arduino, from a simple 8-bit resistor ladder all the way up to using a serial D/A converter. Of these versions, I have found the Microchip MCP4725 12-bit I²C D/A converter shown in **Figure 3.46** to be the simplest and easiest to use.

Figure 3.46 — Adafruit MCP4725 breakout board.

Using the MCP4725 Arduino library and example sketches, you can quickly and easily have your Arduino output analog voltages, sine waves, and triangle waves.

Figure 3.47 — Atmel 24C64 64K I²C EEPROM.

Other Devices

There are number of individual chips and components you may want to use in your Arduino projects. Here are a few that I keep on hand in my parts supply.

As you may recall, some of the Arduinos and variants do not have onboard EEPROM memory. To save data such as calibration settings and other values that you don't want to go away after a reset or power cycle, you can use a serial EEPROM (**Figure 3.47**). Available from Atmel and Microchip in sizes from 128 bit up to 1 megabit, serial EEPROMs communicate with the Arduino via SPI or I²C, are well supported by Arduino libraries and are easy to use.

I²C Digital I/O Expanders

Sometimes, the 14 pins of digital I/O available on the standard Arduino just aren't enough, and you don't want to splurge for a Mega or a Due. Using an 8 or 16 pin I/O expander chip may just fit the bill and only cost you a couple of dollars. The Microchip MCP23008 I²C serial I/O expander chip (**Figure 3.48**) will add eight pins and the Microchip MCP23017 (**Figure 3.49**) will add 16 pins of digital I/O to your Arduino. Fully supported by Arduino program libraries, the serial I/O expander chips communicate with the Arduino using the I²C bus and have programmable address settings, allowing for up to eight devices on a single I²C bus. This means you can add up to 64 or 128 pins of digital I/O to your Arduino to handle the biggest of Arduino projects. The MCP23S08 and MCP23S17 provide similar capability, but communicate using the SPI bus.

Figure 3.48 — Microchip MCP23008 8-bit port expander.

Figure 3.49 — Microchip MCP23017 16-bit port expander.

Analog Switch Chips

Your project may have a need to switch audio or other analog signals but you don't want to waste the power and board space on a relay. For these projects, there are the Analog Devices ADG888 CMOS switch (**Figure 3.50**) and the IXYS LCC120 OptoMOS relays (**Figure 3.51**). The ADG888 is a double-pole, double-throw (DPDT) CMOS chip designed for high performance audio switching and has an ultralow "on" resistance of 0.8 Ω and is capable of switching up to 400 mA at 5 V. The LCC120 is an optically-driven SPDT MOS relay that has an "on" resistance of 20 Ω and is capable of switching 170 mA at 250 V.

Digital Potentiometer Chips

Digital potentiometers allow your Arduino sketches to control the resistance of a potentiometer, just as if you were turning the potentiometer manually. The

Figure 3.50 — Analog Devices ADG888 dual DPDT CMOS switch.

Figure 3.51 — OptoMOS relay.

Microchip MCP42XXX series of digital potentiometers (**Figure 3.52**) are dual 256-step digital potentiometers, available in 10 kΩ, 50 kΩ, and 100 kΩ versions (The last three digits of the chip number is the resistance value). These digital potentiometers communicate with the Arduino using the SPI bus and are well supported by Arduino libraries.

Figure 3.52 — Microchip 8-bit digital potentiometer.

MAX7219 LED Driver

The Maxim MAX7219 (**Figure 3.53**) allows you to drive up to eight 7-segment LEDs (**Figure 3.54**), multiple bar graph LEDs, an 8×8 LED array (**Figure 3.55**), or 64 individual LEDs from a single chip. The MAX7219 communicates with the Arduino using the SPI bus and handles all the decoding, multiplexing, and scan timing for the LEDs on the chip itself.

Figure 3.53 — Maxim MAX7219 serial LED driver.

Other Components

Some other components you will probably want in your parts stockpile include the standard group of infrared (**Figure 3.56**), ultrasonic ranging, proximity, motion, and vibration (**Figure 3.57**) sensors. For motor driving, I like to have some STMicroelectronics L298N H-bridge chips, LG9110h Half-H-Bridge chips, and the Texas Instruments ULN2003AN Darlington array chip on hand. When I need to convert between TTL serial and RS-232, I like to use the Maxim MAX232 dual receiver/driver chip.

In the area of general components, I also recommend a stock of LEDs (red, green, blue, white, RGB, bar graph, 7-segment, and 8×8 array), diodes (silicon, Schottky, Zener, and photo), 2N2222 and 2N3055 transistors for relay driving and other high current needs, photoresistors, capacitors (both ceramic and electrolytic), and 4N25 optoisolators to provide switching isolation between the Arduino and things such as rig-keying circuits.

Figure 3.54 — 7 Segment LED display.

Don't forget an assortment of switches. In addition to the standard mini switches, I like to keep a couple binary coded decimal (BCD) switches (**Figure 3.58**) on hand, to allow me to use one switch and four digital I/O pins to select up to 16 settings. It's also good to have some op amps on hand. I prefer the LM324 dual op-amp, and for power audio applications I like the LM386 audio amplifier chip. No parts bin would be complete without a couple LM555 timers and NE567 PLL tone decoder chips.

For my projects involving Arduino protoshields, I like to use the DuPont-style 2.54 mm-spacing pins and housings, which allow you to easily disconnect the external components as needed.

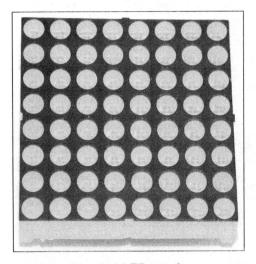

Figure 3.55 — 8×8 LED matrix.

Figure 3.56 — Zilog passive infrared motion detector module.

Figure 3.57 — MEAS vibration sensor.

Figure 3.58 — Hexadecimal binary encoded switches.

Arduino Shields, Modules, and Devices   3-33

Enclosures

Once you've built your project, you have a number of enclosure options available. **Figures 3.59** to **3.70** show some of the wide variety of enclosures available.

There are several enclosures for the Arduino Uno and Mega footprints that allow you stack one shield on the processor board and still fit nicely in the enclosure. This is why I prefer to build my projects with prototyping shields, individual modules, and components. If you use a Mega-size enclosure with an Arduino Uno, you have room for a 9 V battery, display, a module or two, switches, and even a potentiometer allowing you to put a fairly complex project into a nice looking enclosure.

Of course, there is even the ubiquitous Altoids mint tin insert for the Uno. Everyone knows you have to have at least one Altoids mint tin project in your shack.

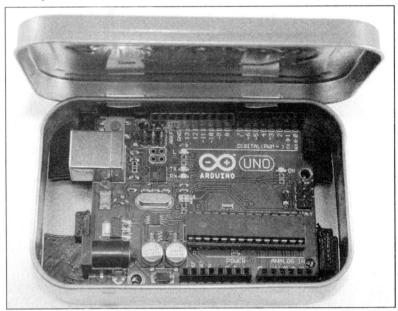

Figure 3.59 — Arduino Uno in Altoids mint tin.

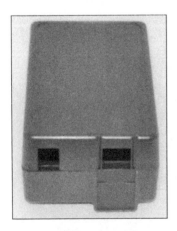

Figure 3.60 — SparkFun project enclosure (product code: 10088).

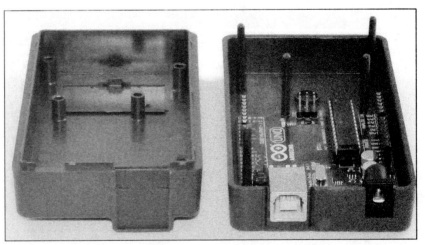

Figure 3.61 — SparkFun project enclosure (product code: 10088).

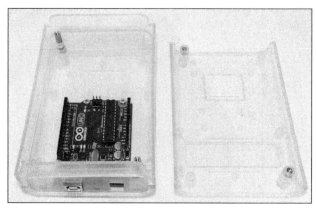

Figure 3.62 — SparkFun clear enclosure for pcDuino/Arduino (product code: PRT-11797).

Figure 3.64 — Acrylic Arduino Mega enclosure.

Figure 3.63 — Acrylic Arduino Uno enclosure.

Figure 3.65 — Solarbotics S.A.F.E. Arduino Uno enclosure (product code: 60100).

Arduino Shields, Modules, and Devices 3-35

Figure 3.66 —Solarbotics S.A.F.E. Arduino Mega enclosure (product code: 60105).

Figure 3.68 — SparkFun Crib enclosure (product code: PRT-10033).

Figure 3.67 — SparkFun Chameleon enclosure (product code: PRT-09682).

Figure 3.69 — Zigo Arduino Uno enclosure.

Figure 3.70 — Zigo Arduino Mega enclosure.

References

4D Systems — www.4dsystems.com.au
Adafruit Industries — www.adafruit.com
AMS — www.ams.com
Analog Devices — www.analog.com
Arduiniana — www.arduiniana.org
Arduino — www.arduino.cc
Atmel Corp — www.atmel.com
Bosch — www.bosch-sensortec.com
Crisp Concept — www.crispconcept.com
D&D Engineering — sensorguys.com
Grand Idea Studio — www.grandideastudio.com
IXYS Integrated Circuits — www.ixysic.com
Maxim Integrated — www.maximintegrated.com
Microchip Technology — www.microchip.com
Midnight Design Solutions — www.midnightdesignsolutions.com
Parallax — www.parallax.com
Pololu Robotics and electronics — www.pololu.com
RadioShack — www.radioshack.com
Skylab Technology Company — www.skylab.com.cn
Solarbotics — www.solarbotics.com
SparkFun Electronics — www.sparkfun.com
STMicroelectronics — www.st.com
Texas Instruments — www.ti.com
Tindie — www.tindie.com
Wikipedia — www.wikipedia.com
ZiGo — www.zigo.am

CHAPTER 4

Arduino I/O Methods

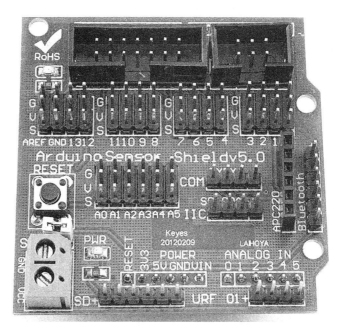

The Arduino Uno sensor shield makes it easy to interface devices and modules to your Arduino.

The ability of the Arduino to interface with so many different sensors and devices is one of the main reasons for its popularity. The primary methods of I/O on the Arduino are digital I/O, digital I/O with pulse width modulation (PWM), analog input, and TTL/USB serial communication. The Arduino supports several bus-type protocols, including the Serial Peripheral Interface (SPI), Inter-Integrated Circuit (I²C), and 1-Wire. The Arduino also supports hardware and software interrupts, which allow program execution flow to be changed (interrupted) by an external event.

Digital I/O

The simplest and most common form of I/O on the Arduino is done using the digital I/O pins. Using the onboard digital I/O pins, the Arduino can turn on LEDs, relays, and other external devices with a single pin as well as sense a switch, pushbutton, or other form of on-off (digital) input. Taken to the next level, there are several program libraries for the Arduino that allow you to use digital I/O to emulate TTL serial or SPI communication on pins other than the pins designated for hardware serial and SPI communications. The Arduino digital I/O pins also have an internal pull-up resistor (normally disabled) that can be enabled via software. On the Arduino Uno, Mega, and Leo the value of the pull-up resistor is 20 kΩ to 50 kΩ. On the Arduino Due, the value of the pull-up resistor is 100 kΩ.

Digital I/O with Pulse Width Modulation

Six of the digital I/O pins (pins 3, 5, 6, 9, 10, and 11) on the Arduino Uno can be configured to output data using pulse width modulation (PWM). PWM allows you to control the duty cycle of a square wave output on the pin using an 8-bit value from 0 to 255. Using PWM, you can dim an LED, output an audio tone, control the speed of a dc motor, modulate a signal for an infrared LED remote, and perform other tasks where you need to modify the duty cycle of the output waveform. The Arduino Leo has seven PWM outputs, the Arduino Mega has 14 PWM outputs, and the Arduino Due has 12 PWM outputs.

Analog Input

The Arduino Uno has six analog input pins. These pins convert an input of 0 to 5 V (3.3 V on the 3.3 V Arduinos) into a 10-bit value of 0-1023. The range can be modified either using an external reference voltage or selecting an internal reference voltage of 1.1 V. The Arduino Mega also supports an internal reference voltage of 2.56 V. The Arduino Leo, Mega, and Due each have 12 analog inputs onboard. The Arduino analog I/O pins also have an internal pull-up resistor (normally disabled) that can be enabled via software. On the Arduino Uno, Mega, and Leo the value of the pull-up resistor is 20 kΩ to 50 kΩ. On the Arduino Due, the value of the pull-up resistor is 100 kΩ.

Serial I/O

The Arduino Integrated Development Environment (IDE) on your PC communicates with the Arduino via the hardware serial port. Most Arduinos communicate with the IDE via the USB connector. However some Arduinos, such as the Solarbotics Ardweeny, require a USB-to-TTL serial converter (FTDI module) to access the serial port. The USB port shares the hardware serial port, located on pins 0 and 1 on the Arduino Uno.

The IDE can be used to upload sketches and also has a Serial Monitor option that allows you to view the serial port data and to send serial data to the Arduino. Because the Uno shares the serial port with the USB port, only one of the two can be active at a time. To get around this limitation, you can use the `Software Serial` program library, which allows you to assign any pair of digital I/O pins as a software-driven serial port.

The Digilent chipKIT Uno32 has two hardware serial ports while the Arduino Mega and Due each have four hardware serial ports onboard.

1-Wire Communications

The 1-Wire Interface was designed by Dallas Semiconductor Corp (now part of Maxim Integrated Products) to provide low-speed data signaling and power over a single data line. Typically used to communicate with small devices such as temperature sensors, voltage sensors, current sensors, memory, and other devices, the 1-Wire interface is actually a bus architecture implemented with a single wire. With a data rate of up to 16.3 Kb/s, the 1-Wire bus can communicate reliably with devices over 100 meters away from the host.

Each 1-Wire device has its own unique 64-bit serial number embedded in the device, allowing many 1-Wire devices to be attached to the same bus without the need to configure the device. Using an advanced algorithm, the host, also known as the bus master, can quickly identify all of the devices attached to the bus and determine their type based on the device type embedded in the lower 8 bits of each device's serial number. This algorithm can scan the 1-Wire bus and identify up to 75 sensors per second.

A unique feature of the 1-Wire interface is that many 1-Wire devices can be powered entirely from the data line and do not need a separate supply voltage to operate. Known as "parasitic power," a small capacitor (typically 800 pF) is integrated within the device to store enough of a charge to operate the device interface.

MaxDetect also makes a series of relative humidity and temperature sensors that use a proprietary 1-Wire interface. The MaxDetect 1-Wire interface is not a bus architecture and the MaxDetect devices do not have embedded addresses, meaning that you can only have one device attached to the interface at a time. The MaxDetect 1-Wire interface does not support parasitic power mode and must be supplied power on a pin separate from the data pin. The MaxDetect 1-Wire interface is not compatible with the Dallas Semiconductor 1-Wire bus, so care must be taken when mixing these devices in your projects.

Both types of 1-Wire interfaces are supported by Arduino example sketches and libraries, which makes interfacing and using 1-Wire devices in your Arduino projects simple and easy.

Serial Peripheral Interface (SPI) Bus

The Serial Peripheral Interface (SPI) protocol was developed by Motorola to be a high speed, full-duplex communication, bus-type protocol between one master and multiple slave devices. The Arduino communicates with the SPI devices on a single shared bus using four signal lines. The signal lines are labeled Clock (SCLK), Slave or Chip Select (SS or CS), Master-Out Slave-In (MOSI), and Master-In Slave Out (MISO). Each SPI device on the SPI bus requires a separate Slave Select line.

SPI is a loosely defined standard and can be implemented in slightly differing ways between device manufacturers. Since SPI is considered to be a synchronous communications protocol, data is transferred using the SCLK line, with no defined upper limit on speed. Some SPI implementations operate at over 10 Mb/s. SPI has four defined modes (Modes 0, 1, 2, 3) which define the SCLK edge on which the MOSI line clocks the data out, the SCLK edge on which the master samples the MISO line, and the SCLK signal polarity. Fortunately, the Arduino program libraries for SPI and the various SPI devices handle the proper signaling required for the various devices.

On the Arduino Uno, the SCLK, MISO, and MOSI pins are permanently defined as digital pins 13, 12, and 11 respectively. Digital I/O pin 10 is often defined as the Slave Select pin for the first SPI device. On the Arduino

Mega2560, the SCLK, MOSI, and MISO are defined as digital I/O pins 52, 51, and 50 respectively, with SS for the first SPI device typically assigned to digital I/O pin 53. On the Uno, Leo, Mega2560, and Due the SPI signals are also brought out to the 6-pin In-Circuit Serial Programming (ICSP) header.

The Arduino LEO and Due do not have any digital I/O pins assigned for SPI. Instead the SPI signals are brought out to the ICSP header only. Since each SPI device attached to your Arduino requires a digital I/O pin assigned to the Slave Select of each SPI device, your projects are limited to the number of digital I/O pins available.

There are several Arduino program libraries available that allow you to software-define the SPI digital I/O pins and communicate with SPI devices on those pins using software-based timing of the SPI signals (also known as bit-banging). With this method, the digital I/O pins are turned on and off manually to simulate the hardware timing needed to communicate with the device. While this method does work, it requires all of the bit timing and clocking to be performed in software, which is much less efficient than the hardware-based SPI method. It is far better to design your projects to use hardware SPI and use software SPI communication only when required.

Inter-Integrated Circuit (I²C) Bus

The Inter-Integrated Circuit (I²C) bus was developed by Phillips (now NXP Semiconductors) for attaching low-speed peripherals to a host device. On the Arduino, the I²C bus is also known as the Two-Wire Interface (TWI) bus. I²C is a serial bidirectional 8-bit communication protocol used by many manufacturers developing peripherals and devices for embedded systems such as the Arduino.

I²C requires only two communication lines, Serial Data (SDA) and Serial Clock (SCL). On the Arduino Uno, these are defined as analog pins A4 and A5 respectively. On the Arduino Leo, SDA and SCL are assigned to digital pins 2 and 3 respectively. On the Arduino Mega2560 and Due, SDA and SCL are assigned to pins 20 and 21 respectively. The Arduino Due has a second I²C interface, assigned to pins SDA1 and SCL1.

The I²C standard defines the speed of the I²C bus as 100 Kb/s (Standard Mode), 400Kb/s (Fast Mode), and 3.4 Mb/s (High Speed Mode). The Arduino defaults to an I²C bus speed of 100 Kb/s, but the speed can be changed by modifying the internal Arduino Two Wire Bit Rate Register (TWBR) or the TWI speed definition in the Arduino `Wire` library. Unless your project requires it, it is best to not modify the `Wire` library, as it may cause issues when you compile other projects using the same `Wire` library.

I²C devices have a unique 7-bit address, with some devices capable of having their I²C address reassigned using jumpers, switches, or some other hardware method, allowing you to have multiple devices of the same type co-exist on the I²C bus. On the Arduino, I²C addresses 0-7 and 120-127 are reserved, leaving 112 addresses available for devices. Every I²C device connects to the bus using open-drain outputs, requiring pull-up resistors on the bus, typically 4.7 kΩ.

As with software SPI, there are several Arduino program libraries available that allow you to software-define the I²C digital I/O pins and communicate with I²C devices on those pins using software-based timing of the I²C signals (also known as bit-banging). With this method, the digital I/O pins are turned on and off manually to simulate the hardware timing needed to communicate with the device. While this method does work, it requires all of the bit timing and clocking to be performed in software, which is much less efficient than the hardware-based I²C method. It is far better to design your projects to use hardware I²C and use software I²C communication only when required.

Interrupts

While not a true I/O method, interrupts allow external conditions and events to modify the way your sketches execute. The Arduino has two types of interrupts — hardware and timer. Hardware interrupts are triggered by an external event, such as a change in voltage level on a digital I/O pin. An interrupt will pause the current program execution and execute a user-defined function, known as an interrupt handler or interrupt service routine (ISR). When the ISR is complete, program execution resumes the program at the command that was executing prior to the interrupt. An interrupt can happen at any time and allow you to respond to external events without constantly checking to see if the desired condition exists.

There are four types of interrupt conditions that can be defined: Rising, Falling, Change, and Low. These interrupts conditions refer to the state of the digital I/O pin. The Rising condition will generate an interrupt when the pin goes from a low state to a high state; Falling will generate an interrupt when the pin goes from high to low. Change will generate an interrupt when the pin changes from either low to high or high to low. The Low condition will generate an interrupt when the pin is low. The Arduino Due has an additional interrupt condition, High, which generates an interrupt when the pin is high.

The Arduino Uno has two interrupts, assigned to digital I/O pins 2 and 3. The Arduino Leo has four interrupts, on pins 0, 2, 3, and 7, the Arduino Mega has six interrupts, on pins 2 and 3 and 18 to 21. The Arduino Due allows you to configure interrupts on all available pins. The Arduino Uno can also handle Change interrupts on all I/O pins, but unlike the hardware interrupts, the ISR function must decode the interrupt and determine which pin generated the interrupt.

The Arduino Uno has three internal timers, defined as `Timer0`, `Timer1`, and `Timer2`, which can be used to generate `Timer` interrupts. `Timer0` is an 8-bit timer which is used by the Arduino for internal timing functions such as `delay()` and `millis()`. Since it can affect these functions, modifying `Timer0` settings is not recommended. `Timer1` is a 16-bit timer which is used by the Arduino `Servo` library, and `Timer2` is an 8-bit timer used by the Arduino `tone()` function. As long as you are aware of any potential interaction with these functions, you can modify the settings on `Timer1` and `Timer2` for use in your sketches. The Mega2560 has three additional 16-bit timers — `Timer3`, `Timer4`, and `Timer5` — which are not used by any Arduino internal functions.

You can configure the timers to generate a software interrupt on overflow or when the timer count matches a desired value. The timers are based on the

Arduino CPU clock rate, typically 16 MHz. You can use the timer counter/control register (TCCR) for each timer to control the timer clock setting. By modifying the three Clock Select bits, you can control how fast the timer increments the counter. Available settings are Clk/1 (Clock speed), Clk/8, Clk/64, Clk/256, and Clk/1024. At the maximum setting of Clk/1024, you can have 16-bit `Timer1` generate a software interrupt approximately every 4.194 seconds. Using the Clear Timer on compare match (CTC) setting, you can adjust the timer to generate an interrupt when the timer reaches a preset value. Using 15624 as the preset value will cause the timer to generate an interrupt once per second. Using a 1-second interrupt in this manner, you can add precision timing to your sketches without having your sketches manually keep track of time using the `millis()` function or other manual methods.

Implementing interrupts does add a level of complexity to your sketches, and due to their asynchronous nature, interrupts can occur at any time. You will have to remember and plan for this as you write and troubleshoot your sketches. You can enable and disable the interrupts as needed within your sketches, to allow for critical points where you don't want your main sketch loop to be interrupted.

Used properly, interrupts can be a powerful tool in developing your Arduino projects and once you get comfortable using them, you will find that interrupts can help simplify your project development since you will no longer have your sketches running in timing loops waiting for an event to occur. Instead, you can have your sketch doing other things and only respond when the actual event occurs.

References

1-Wire.org — **www.1wire.org**
Arduino — **www.arduino.cc**
D&D Engineering — **sensorguys.com**
Diligent — **www.digilent.com**
Maxim Integrated — **www.maximintegrated.com**
NXP Semiconductor — **www.nxp.com**
Solarbotics — **www.solarbotics.com**

CHAPTER 5

Arduino Development Environment

Arduino programs, or sketches, are developed using the Arduino Integrated Development Environment (IDE) running on a workstation. Derived from the Processing Programming Language and Wiring projects, the Arduino IDE is a cross-platform development tool that runs on *Windows* PC, Mac *OS X*, or *Linux* workstations. There's even an IDE for the Arduino that runs on Android devices. The Arduino IDE is used to compose sketches, connect to and upload sketches to the Arduino, and to communicate with your sketches. The IDE is available for download from the Arduino website (**arduino.cc/en/main/software**).

The Arduino IDE includes a text editor to create your sketches. You can select from more than 30 languages and even use your favorite external text editor by changing the IDE preference settings. The Arduino text editor includes syntax and keyword highlighting, brace and bracket matching to identify program loops and if-then coding blocks, and it automatically indents your program text based on the depth of the current program loop. Additional

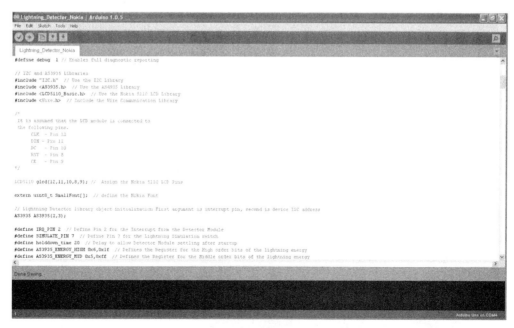

Figure 5.1 — Sample screen from the Arduino Integrated Development Environment.

preferences can be set in the preferences.txt file located in **c:\Documents and Settings\<USERNAME>\Application Data\Arduino** for *Windows XP*, **c:\Users\<USERNAME>\AppData\Roaming\Arduino** for *Windows Vista* and *Windows 7*, **/Users/<USERNAME>/Library/Arduino/** for Mac *OS X*, and **~/.arduino/** for *Linux*. Only edit the preferences.txt file when the Arduino IDE is not running as the file is re-written every time you exit the IDE. **Figure 5.1** shows a sample IDE screen.

The Arduino IDE does more than just create sketches. The IDE has a message area to provide feedback while saving and exporting sketches, and it is where any error messages are displayed. The toolbar provides quick access to the most commonly used IDE functions. The toolbar VERIFY function checks your sketch for basic errors, but does not upload the sketch to the Arduino. The toolbar UPLOAD function will verify, compile, and upload your sketch to the Arduino board. It is important to note that the Arduino IDE performs an auto-reset on the Arduino board prior to uploading. Some older Arduino boards, along with some of the variant boards, do not support the auto-reset function. On these boards you will have to manually press the reset button immediately prior to selecting UPLOAD in the IDE. The remainder of the IDE toolbar displays the NEW, OPEN, and SAVE sketch file functions.

Menus

The FILE menu of the IDE is the usual area where you can create new sketches, load and save sketches, print sketches, and set your IDE preferences. The FILE menu also has a quick-access dropdown menu that lists the programs in your sketchbook (the folder where the IDE stores your Arduino projects). Another quick-access dropdown menu provides quick access to example sketches that are typically provided with the various libraries installed in the IDE. At the time this was written in early 2014, these dropdown menus don't have scrolling capability due to a bug between the IDE and Java. There is an enhanced version of the IDE in the Arduino forum (**http://forum.arduino.cc/index.php?topic=118440.0**) that you can download and install, or you can use the standard FILE/OPEN menu to select the desired sketch.

In addition to standard editing menu features, the EDIT menu allows you to copy your sketch to your workstation's clipboard and reformat the text, including the syntax coloring, in a form suitable for posting in the Arduino forums and in an HTML format suitable for posting on web pages.

The SKETCH menu allows you to verify and compile your sketch as well as providing a quick method to display the contents of the sketch folder on your workstation. The SKETCH menu also allows you to add other code into your project, which appears on separate tabs in the IDE text editing area. This allows you to easily merge pre-existing sketches and code into your current project. These added files can be Arduino code, C code (.c files), C++ code (.cpp files), or header files (.h files). Also on the SKETCH menu is the IMPORT LIBRARY function, which will install the selected library into the IDE and add the `#include` declaration for the selected library at the start of your sketch. Starting with IDE Version 1.0.5, you can also import libraries in .ZIP format.

The TOOLS menu is where you select the type of Arduino board you will be programming and which serial port the board is attached to. If you don't set these to the proper board and port, your sketch will verify and compile, but the upload will fail. You can also add support for third-party hardware (such as variant boards) by placing their board definitions, core libraries, bootloaders, and programmer definitions into a sub-folder in the hardware folder in your sketchbook. You can create the hardware folder if it does not exist and the IDE will incorporate the new board into the IDE menus the next time the IDE starts.

A handy feature on the TOOLS menu is the AUTO FORMAT option. This will reformat the text in your sketch so that the curly braces all line up and the text within the braces is properly indented, making your programs easier to read, which will help when troubleshooting your sketches. Another helpful utility on the TOOLS menu is the ARCHIVE SKETCH option. This will create a copy of your current sketch and save it in the current sketch folder as a compressed .ZIP file.

Also on the TOOLS menu you will find options to select the hardware programmer used to upload and save your sketch. You will usually only need to change this option when you are programming a board or chip that does not use an onboard USB/serial port. This setting is typically used when burning the Arduino bootloader. Since most of the Arduino boards you will be working with already have a bootloader, the only time you will have to burn the bootloader is if you are trying to recover an Arduino that has lost its bootloader or you are programming a chip that does not already have a bootloader installed.

One of the most important options on the TOOLS menu is the SERIAL MONITOR. The SERIAL MONITOR will display serial data sent from the Arduino. Typically the SERIAL MONITOR is used to show basic program output and provide debugging information. You can also use the SERIAL MONITOR to send characters and commands to the Arduino serial port.

Arduino Libraries

Libraries are what make the Arduino so easy to program. Many of the shields and devices you can attach to the Arduino are supported by program libraries. By using an existing library, a large percentage of the programming needed for a project is already done for you. Think of libraries as the core building blocks of your Arduino sketches. All you have to add is a little "glue" to tie the libraries into your project and you're done. This will save you hours and days of deciphering datasheets and writing test code just to interface to a new device. Now, all you have to do is get the library for the device, look at the example code that comes with most libraries and start writing your sketch. In many cases, you can cut and paste parts of the example sketch into your sketch and have a completed, working project in a matter of hours.

There are many "official" libraries that come integrated with the Arduino IDE when you install it, as well as a large number of "unofficial" libraries, created and shared by the vast community of Arduino developers. Herein lies the true power of the Open Source community. Because the Arduino is Open Source, there are many hobbyists just like yourself out there, creating projects with the same devices you are. Many of these hobbyists freely share their projects and libraries, saving you from having to create your own. Since everything is Open Source, you can take someone else's library and modify it further to suit your specific needs. In most cases, you will find that an existing library supports all the functions of the device that you will need in your project.

Internal Libraries

The Arduino IDE includes 16 libraries when it is installed. These libraries are:

EEPROM — contains a set of functions for reading and writing data to the Arduino's internal EEPROM memory.

Esplora — contains a set of functions for interfacing with the sensors and switches on the Arduino Esplora board.

Ethernet — contains a set of functions for using an Arduino with an Ethernet shield to connect to the Internet. The library allows you to configure the Arduino to function as either a web server or as a web client.

Firmata — contains a set of functions to implement the Firmata communications protocol for communicating between Arduino sketches and the host computer.

GSM — contains a set of functions for using an Arduino with a GSM shield. This allows the Arduino to act as a GSM phone, placing and receiving calls and SMS messages, and connecting to the Internet over the GPRS network.

LiquidCrystal — contains a set of functions for communicating with Hitachi HD44780-compatible parallel LCD displays.

Robot_Control — contains a set of functions to interface with peripherals and devices on the Arduino robot control board.

Robot_Motor — contains a set of functions to interface with the Arduino motor control board.

SD — contains a set of functions for reading and writing SD memory cards.

Servo — contains a set of functions to control servo motors.

SoftwareSerial — contains a set of functions to allow the use of serial communications on I/O pins other than those normally used for serial communication.

SPI — contains a set of functions used to communicate with SPI devices.

Stepper — contains a set of functions to control unipolar and bipolar stepper motors.

TFT — contains a set of functions to draw shapes, lines, images, and text on a thin-film-transistor (TFT) display.

WiFi — contains a set of functions to allow an Arduino with a WiFi shield to connect to the Internet. The library allows you to configure the Arduino to function wirelessly as either a web server or as a web client.

Wire — contains a set of functions to communicate with I^2C devices.

Contributed Libraries

These are also a number of contributed libraries available on the Arduino Playground website (**http://playground.arduino.cc**). The Playground is where Arduino users can gather and share their thoughts, ideas, suggestions, and Arduino code. There is a huge amount of information on the Playground Wiki, including tutorials, how-tos, tips, tools, and libraries. Some of the contributed libraries you may want to use in your projects are:

Adafruit_GFX — contains a set of functions to draw shapes, lines, images, and text on Adafruit graphics displays.

AS3935 — contains a set of functions for communicating with the AS3935 Franklin lightning detector.

Bounce — used to debounce digital inputs.

CapacitiveSensor — allows you to turn two Arduino I/O pins into a capacitive touch sensor.

FFT — contains a set of math functions allowing you to perform fast Fourier transform (FFT) operations on your data.

Flash — contains a set of functions to store static strings, arrays, tables, and string arrays in flash memory.

GLCD — contains a set of functions to support a large number of graphic displays, including functions to draw shapes, lines, images, and text.

LCD_I2C — contains a set of functions for communicating with Hitachi HD44780-compatible I²C LCD displays.

Messenger — used to parse serial text sent to the Arduino and place the received data into an array.

Morse — generates Morse code from standard text.

OneWire — used to communicate with devices from Dallas Semiconductor (now Maxim Integrated) using the 1-Wire protocol.

PCD8544 — contains a set of functions to draw shapes, lines, images, and text on the Nokia 5110 LCD display.

PString — contains a set of functions to format text for output.

PS2Keyboard — used to interface a PS/2 keyboard to the Arduino.

Streaming — provides a set of functions to implement Java/VB/C#/C++ style concatenation and streaming operations.

Time — contains a set of timekeeping functions to allow your sketches to keep track of time and dates with or without external hardware timekeeping.

Timer — used to create and manage software `Timer` interrupts.

TinyGPS and TinyGPS++ — provides a set of functions for communicating with a serial GPS module.

Webduino — allows you to configure the Arduino with a Wiznet-based Ethernet shield to function as an extensible web server.

X10 — enables communication with X10 home automation devices.

Xbee — enables communication with Xbee wireless devices.

Arduino Due Libraries

There also several libraries specific for the Arduino Due:

Audio — enables playback of .WAV audio files.

Scheduler — contains a set of functions to allow the Arduino Due to run multiple functions at the same time without interrupting each other.

USBHost — enables communication with peripherals such as USB mice and keyboards.

Creating Arduino Projects

Now that you know what the Arduino can do, how do you get started? By now you've probably copied an existing sketch or run some of the examples, but how do you create your own projects from scratch?

The first step is to sit back and map out exactly what your project should do. Way back in the dear dead mainframe days, we flowcharted our programs before we started doing any actual coding. Flowcharting is nothing more than creating a block diagram for your programs. While I don't do any formal flowcharting with all of the "official" flowcharting symbols when creating my sketches, I do draw out a basic block diagram of everything my sketch will do. I also draw out a schematic for any circuits I will need to attach to my Arduino. Doing this helps me to break the project into smaller building blocks while keeping track of everything the sketch needs to do.

Doing all of this ahead of time helps you to see if there is anything you have left out or if you have made mistakes in the design, allowing you to correct any issues before you get too far along into your project. It's no fun to have to completely start a project over because you left out a critical function that can't easily be corrected, or you've just miswired that $20 chip and let all the smoke out.

The flowcharting process helps to organize your thoughts and provides a reference point as you create you sketch. Done right, a flowchart such as the one in **Figure 5.2** will help break the sketch out into bite-size pieces that are much easier to code and troubleshoot. With flowcharts you can determine what variables and variable types you will need, what libraries your sketch will need, and what I/O pins will be used. You can determine if a part of the sketch should be a function rather than part of the main loop, or if the block of code would perform better if it used interrupts.

Schematic drawings help to do the same thing that flowcharts do, only for hardware. Taking the time to draw a schematic like the example in **Figure 5.3** allows you to see how all the parts interconnect and what I/O pins you need to connect on the Arduino. Then, when constructing your project, you can refer

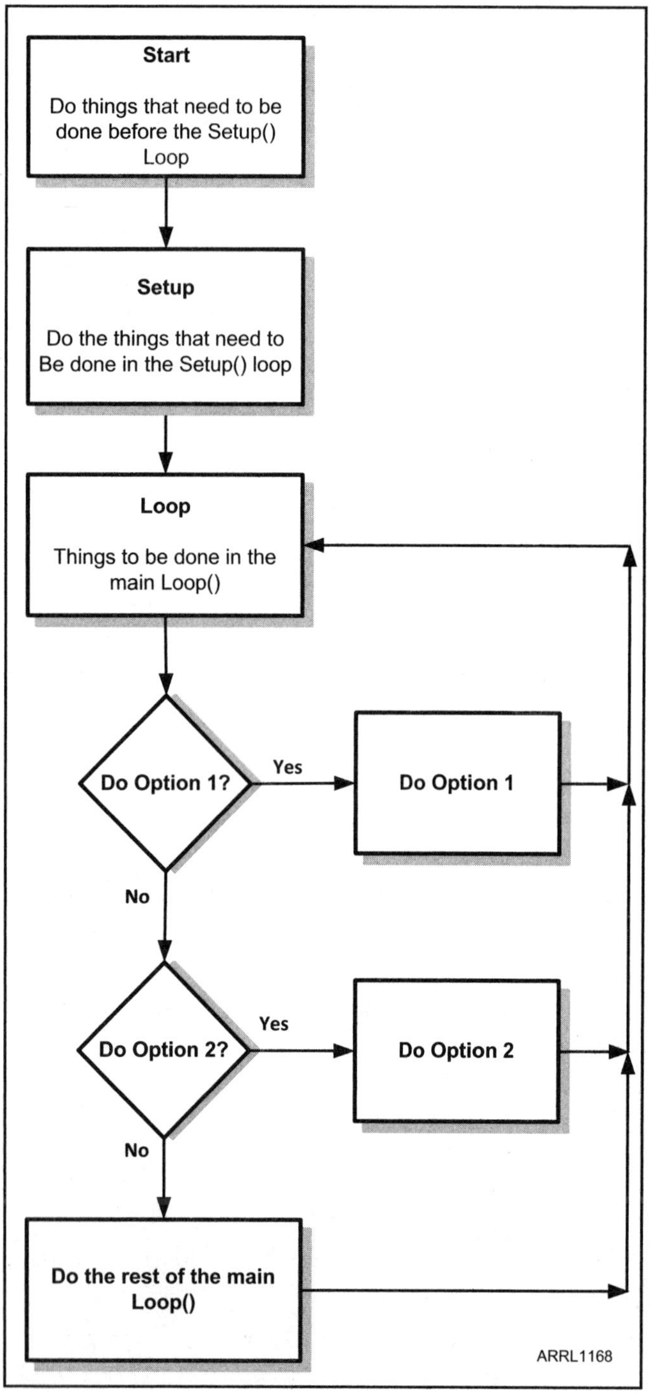

Figure 5.2 — Flowchart example.

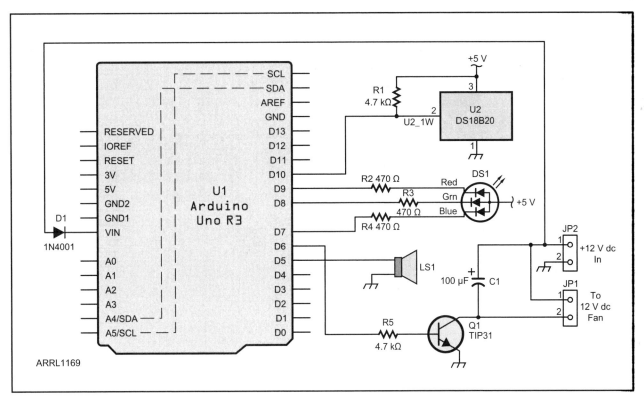

Figure 5.3 — Schematic diagram example.

to the schematic and follow it step-by-step. This helps prevent wiring errors and helps you to find any circuit design errors you may have overlooked. Having a schematic has saved me more than once, as I was one wire away from hooking 12 V to a 5 V A/D input before a recheck of the schematic diagram showed me the errors of my ways. I'm sure that $15 A/D module was happy I stopped to draw a schematic.

You don't have to go into every detail, and you don't have to go so far as to draw a full schematic with every pin labeled (although it does help). Instead, you can just use block diagrams. You know an I²C device will always need the power, ground, SDA and SCL lines connected, so you can just draw one line to represent those items. As you build more projects and become familiar with the components you often use, you can use this form of shorthand to streamline your creative process.

If you want to use a more formal method to create your circuit diagrams, you can use the free version of CadSoft's *EaglePCB* software, or the Open Source *Fritzing* program. Both allow you to create schematic drawings, board layouts, parts lists, and can even generate a file you can use to have your project commercially etched onto a circuit board. Fritzing has a unique feature that allows you to create a breadboard image such as the one shown in **Figure 5.4**, which will automatically draw a schematic that you can use to wire your circuit on an actual breadboard. All you have to do is match up the breadboard image

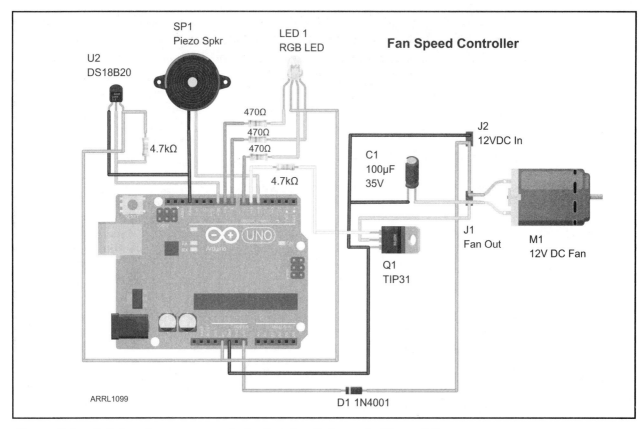

Figure 5.4 — Fritzing diagram example for the circuit shown in Figure 5.3.

to your real breadboard circuit. When you're ready to start soldering, just print the schematic image and you're good to go.

The method you choose to plan your projects is a matter of personal choice, so use the method that works best for you. The key is to take the time to plan everything out ahead before you begin constructing your project or coding your sketch. A little bit of planning and documentation upfront can help keep you from soldering or coding yourself into a corner and getting frustrated. Remember, this is supposed to be fun. Using flowcharts and schematics can help keep things organized and fun.

Memory Management Techniques

In many ways, programming with the Arduino is similar to programming with the early microcomputers such as the 8080, Z-80, and 6502. Memory is at a premium and when it's all used up, you're pretty much out of luck. When you run out of memory in an Arduino sketch, there isn't the handy "Out of Memory" error; instead your sketch just does strange things with no rhyme or reason. You have to always keep in mind that 2 KB of RAM goes in a hurry when you use a lot of strings and arrays.

Fortunately, with the Arduino, you can store static variables and constants in the larger flash memory that the Arduino uses to store your compiled sketches. This can be done using the PROGMEM keyword and including the avr/

`pgmspace.h` library in your sketch. The `pgmspace` library allows you to store static variables, constants, strings, and arrays in flash memory instead of RAM. You can also use the `FLASH` library to store string, array, table, and string arrays in flash. Finally, you can also use the `F()` syntax for storing string constants in flash. Remember, if it goes into flash, it cannot be modified during program execution, so you can only use these functions for data that doesn't change.

Keeping Track of Memory

Fortunately, there are ways you can keep track of the memory your sketches are using. The `MemoryFree` library contains a set of functions that you can use to display the amount of available RAM remaining. By including this library in your sketch, you can experiment and see which coding method you use results in the best utilization of RAM memory.

Simple Debugging Methods

One of the easiest ways I have found to troubleshoot sketches is to include debug code that will print variables and other information as the sketch executes. At the beginning of the sketch, I define a debug preprocessor directive such as `#define DEBUG` and throughout my sketch, I check to see if the debug value is defined using `#ifdef` and `#endif` statements. If my debug value is defined, then I execute a small block of code or a function to print the relevant data to the Serial Monitor. Then, when my sketch is debugged and operational, all I have to do is comment out my debug preprocessor directive and the compiler will ignore my debug print commands, automatically removing all of the debug statements from the uploaded sketch, thereby saving valuable memory space.

References

Arduiniana — **www.arduiniana.org**
Arduino — **www.arduino.cc**
Arduino Playground — **playground.arduino.cc**
CadSoft — **www.cadsoftusa.com**
Firmata — **www.firmata.org**
Fritzing — **www.fritzing.org**
Smarthome — **www.smarthome.com**
X10 — **www.x10.com**

CHAPTER 6

Arduino Development Station

As you begin to create your own Arduino projects, you will quickly find that connecting all the pieces and parts together with jumper wires on a table top doesn't quite do the job. It's all too easy for something to move, fall on the floor, or short out and fry something. In true obedience to Murphy's Law, that fried component will be the most expensive or irreplaceable part in your entire project. Fortunately, the Arduino's size makes it ideal to create a mini-development station that can be set aside without having to tear apart your half-built projects in between development sessions.

Using a small piece of wood, you can turn your Arduino development environment into a stable, portable platform that can be moved and put away in between development sessions without having to take it all apart. **Figure 6.1** shows my Arduino development station. Mounted on a 12 inch by 15 inch board, I have included everything I use on a regular basis when developing my projects.

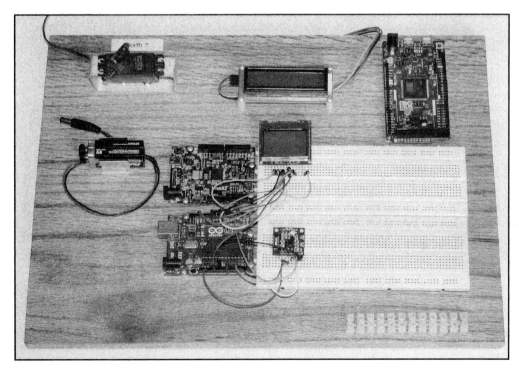

Figure 6.1 — The author's Arduino breadboard/development station.

On my development board I have mounted an Arduino Uno, a chipKIT Uno32, and an Arduino Due. Next to the Arduinos, I have mounted two sections of 7 × 2.5 inch breadboard to use as a circuit design area. Also mounted on the board are brackets to hold a 9 V battery and a servo. At the top of the board, I have mounted a 16-character by 2-line serial I^{2}C LCD that I use in many of my projects. Along the bottom edge of the board, I have a terminal strip to connect wiring for motors and other devices too large to fit on the development board.

Typically, when I begin a project I wire it on a breadboard to create and test my basic designs before moving my project to a more permanent solution. This development board allows me to quickly wire and test my designs, move wires to correct my mistakes, and when I am done, I can just put the development board aside and start soldering up the finished design onto a protoshield.

When I am breadboarding larger projects, I often use an I/O expansion shield on the Arduino, which allows easy access to all of the Arduino pins for connecting jumpers between the Arduino and the circuit on the breadboard.

For smaller and less extensive projects, you can use a breadboard shield. As shown in **Figure 6.2**, this shield gives you a small breadboard area you can use to quickly wire up your project similar to the larger development station, without taking up nearly the space.

Figure 6.2 — Arduino breadboard shield.

When you have completed and tested your prototype project on the breadboard, you may want to move it to a more permanent solution. When using an Uno or similar Arduino, I like to use a prototyping shield (**Figure 6.3**) to mount my components and solder the finished creation. This allows me to permanently mount all of a project's components and easily remove the shield to make any design changes or circuit modifications. For external components, I use the DuPont-style 2.54 mm-spaced pin headers and sockets to connect

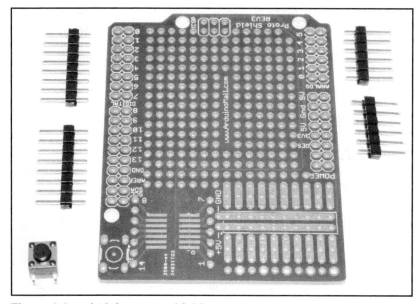

Figure 6.3 — Arduino protoshield.

Figure 6.4 — Hot air soldering station

everything together. This allows me to quickly disconnect and reconnect everything when making changes to the circuits on the prototyping shield.

Soldering Tools

For most of my life, I have used the old standard soldering pencil, which has served me well over the years. When I tried my hand at soldering surface-mount components, I failed miserably with the soldering pencil and borrowed a soldering station with a hot air gun. While my first attempt at surface-mount soldering with the hot air gun is not something that will ever be seen in public, it became obvious that a temperature-controlled soldering station with a hot air gun attachment was the way to go. For about $75, I bought a temperature-controlled soldering iron station (**Figure 6.4**) with a hot air gun for whenever

Figure 6.5 — Digital multimeter.

I get brave enough to make another attempt at soldering surface-mount components. The iron has interchangeable tips, allowing me to quickly switch between tiny circuit board work and welding coax connectors. Don't get impatient like me; be sure to allow the soldering iron to cool before changing tips. The hot air gun also comes with interchangeable nozzles, allowing you to vary the size and pattern of the hot air. The hot air gun also does great work on heat-shrink tubing and testing temperature sensors.

Test Equipment

Regardless of how you create your projects, at some point you will need to troubleshoot them. You don't have to spend a ton of money on test equipment, but there is some basic equipment you will want to have as you start creating your own circuit designs.

Primarily, you will want a digital multimeter (DMM) such as the one shown in **Figure 6.5**, which allows you to read volts, ohms, amps, and check circuit continuity. It doesn't have to be a high-end expensive meter, as most of the time you're just checking for zero, 3.3, or 5 V on an I/O pin, or checking the value of a resistor before soldering it into place. Since they added that extra color band on resistors and messed up everything I ever knew about resistors, I'm finding that I constantly have to check with the meter to be sure I grabbed the right resistor.

As you move into more complex projects, you may find that an oscilloscope (**Figure 6.6**) comes in really handy. An oscilloscope will allow you to see signals on a visual display, usually in the form of a two-dimensional graph, with the vertical axis displaying the voltage and the horizontal axis displaying a function of time. This allows you to view actual waveforms and see signal transitions. Many of the older oscilloscopes used a high persistence cathode ray tube (CRT) to display the signal traces. While many of these older oscilloscopes are still available at hamfests or online for $200 or thereabouts, unless you're like me and just love using them, you really don't need such a serious piece of test equipment to get the job done.

There are a number of USB oscilloscopes (**Figure 6.7**) that can be used with a PC and will do nicely for what you'll be doing. Some cost less than $100. You can even build your own oscilloscope with an Arduino using one of the many projects online, plus you get the added satisfaction of using an Arduino project to troubleshoot another Arduino project.

The difference between the PC-based oscilloscopes and the commercial standalone oscilloscopes is primarily their upper frequency limit, or bandwidth. The old standby Tektronix 465 has a bandwidth of 100 MHz, meaning it can display signals up to 100 MHz, and the Tektronix 475 has a bandwidth of 200 MHz. As you may have guessed, the higher the bandwidth, the more expensive they tend to get. The typical USB oscilloscope connected to your PC has a bandwidth of about 1 MHz, while most Arduino-based oscilloscopes top out around 200 kHz. Since the majority of your troubleshooting will be with much slower signals, any of these oscilloscopes will do the job nicely.

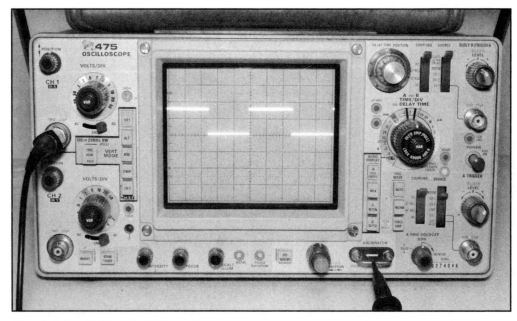

Figure 6.6 — Tektronix 475 oscilloscope.

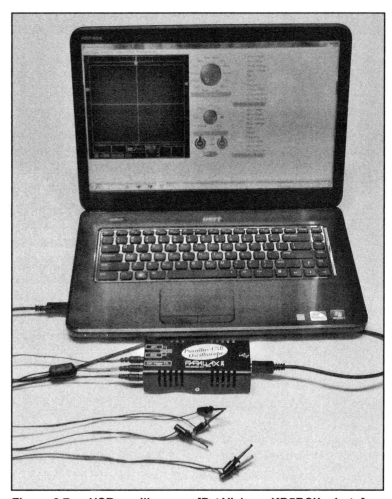

Figure 6.7 — USB oscilloscope [Pat Vickers, KD5RCX, photo]

References

Arduino Playground — **playground.arduino.cc**
Diligent — **www.digilent.com**
Instructables — **www.instructables.com**

CHAPTER 7

Random Code Practice Generator

The finished Random Code Practice Oscillator mounted in a SparkFun pcDuino/Arduino enclosure.

Morse code, or CW as it is commonly called, has always been my first love. My first rig back in my Novice days was a CW-only Heathkit HW-16. It didn't have the external VFO, just a handful of crystals for 15 and 40 meters. All I had for an antenna was a two-element triband beam for 20, 15, and 10 meters, so 40 meters was pretty much out of the question. With my three 15 meter crystals, I hammered out many CW QSOs and had a blast. When I finally got my General license, I discovered the joy of 2 meters and the local 2 meter RTTY gang, so my CW skills faded into near uselessness in favor of the fancy digital modes.

I've never lost my love for CW though, and keep promising myself I'm going to get my act together one of these years and dive back in. The first step of course, is to relearn all that I had forgotten. An Arduino-based Code Practice Generator seemed like just the thing to get me back on track and use the project to get familiar with the Arduino.

The first step was to sit down and flowchart what I wanted my sketch to do (**Figure 7.1**). Once I had created my flowchart, the next step was to draw out a quick block diagram (**Figure 7.2**) that I could use to create the schematic diagram I would use to build the project.

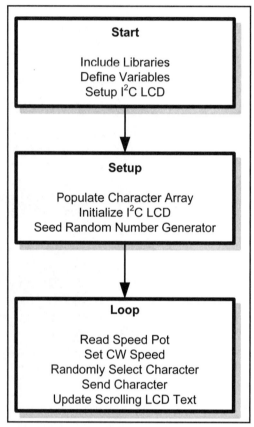

Figure 7.1 — Random Code Practice Oscillator flowchart.

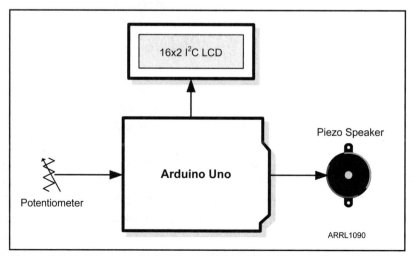

Figure 7.2 — Random Code Practice Oscillator block diagram.

Using the Open Source *Fritzing* design tool, I created a visual representation of how the project would be laid out on my breadboard (**Figure 7.3**). One of the things I like about *Fritzing* is the way that you can visually create your design as it would appear in real life, and *Fritzing* will draw the schematic for you. Another handy feature of *Fritzing* is that it can create a Bill of Materials from your finished diagram, so you'll know what parts you'll need to create your project without having to keep track of all the individual parts manually.

Once I had created my *Fritzing* diagram, I printed out the breadboard diagram and I was ready to begin creating my project. While it takes a while to get used to laying out a project with *Fritzing* so things make sense, and you may have to design some of your own icons for parts not in the *Fritzing* component library (it's not that hard to do), *Fritzing* is a great tool to create your breadboard diagrams. All you have to do is match your breadboard circuit up to the diagram and your circuit is ready for testing. **Figure 7.4** shows the schematic diagram and parts list.

For this project, I wanted the Arduino to randomly select a character to

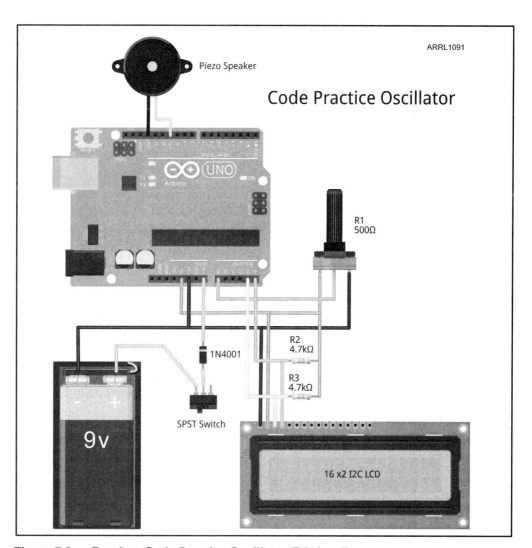

Figure 7.3 — Random Code Practice Oscillator *Fritzing* diagram.

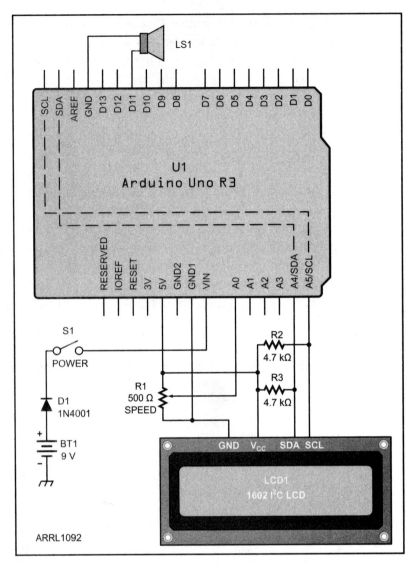

Figure 7.4 — Random Code Practice Oscillator schematic diagram.
BT1 — 9 V battery.
D1 — 1N4001 diode.
LS1 — Piezo buzzer (Radio Shack 273-073 or equiv).
S1 — SPST toggle switch.
R1 — 500 Ω potentiometer.
R2, R3 — 4.7 kΩ, ⅛ W resistor.
U1 — Arduino Uno.
U2 — 16-character × 2-line I²C LCD display.
SparkFun pcDuino/Arduino Project Enclosure, P/N: PRT-11797.
SparkFun pcDuino/Arduino Project Enclosure Extension Plate, P/N: PRT-11798.

send, use an Arduino library to convert the character into Morse code and play it as a tone on a speaker. I also wanted to be able to control the sending speed and to display the sent characters on an LCD display for visual reinforcement of what I was hearing. A brief Internet search provided me with the Arduino `Morse` library by Erik Linder, SMØRVV, and Mark VandeWettering, K6HX. This library is simple and easy to use. All you have to do is include the library in your sketch, define the digital I/O pin you wish to use, the sending speed, and whether you want the library to send CW using a tone or a voltage that can be used to key a relay or transceiver. Once setup, all you do is call the library's `send` function with the desired ASCII character and it will output the Morse code on the assigned pin.

For simple displays, I prefer the I²C version of the Hitachi HD44780-compatible 16-character by 2-line (16×2) display. This display is one of the least expensive and most commonly used LCDs for the Arduino. I prefer the I²C version because it uses the SDA (A4) and SCL (A5) pins and doesn't tie up any of the standard digital I/O pins. Since most of my projects don't need all six analog input pins, and the standard version of the LCD needs six digital I/O pins to communicate, the I²C version allows me the flexibility to assign the Arduino digital I/O pins to other purposes without sacrificing the simplicity and ease of use of this display. The Arduino Playground has the `LiquidCrystal_I2C` library to support the I²C version of this LCD. All you have to do is include this library in your sketch along with the built-in `Wire` library to support I²C communication and you're ready to roll.

Once I had figured out all the pieces I would need to create the sketch, and the electronics I would need to support the sketch, I began laying out the project on the breadboard for testing. What I love about most Arduino projects is that the actual wiring is usually simple and easy. The Arduino and libraries do most of the heavy lifting for you. The key is to have your concept clearly laid out ahead of time and the rest should go easy.

Creating the Sketch

Once you have the circuit wired up on the breadboard, you can begin to create your sketch. Rather than print the entire sketch here, the completed sketch and libraries for this project can be found in **Appendix A**, and are also available for download from **www.w5obm.us/Arduino**.

Using the flowchart as a guideline, you can break the sketch out into smaller building blocks, testing each block as you go along. At the top of the sketch, you will find the usual library `#include` statements, variable definitions and the object assignments for the LCD display. There are two key variables defined in this section, the character array `Code[]`, and the string variable `text`. The purpose of the `Code[]` array is to store the ASCII representation of the numbers, symbols and characters we want our practice oscillator to send. Placing the characters to send in an array allows us to use the Arduino's random number generator function to randomly select the array element that contains the character to send next.

As I began working with the LCD, I discovered an issue with scrolling the LCD text. Many of the libraries for the Hitachi HD44780-compatible displays do not scroll the text on the display correctly and require both lines to scroll simultaneously. Since I only wanted the top line to scroll and the current sending speed to be displayed on the second line without scrolling, I had to create my own scrolling function. The `text` variable is used to contain the top line of information that will be scrolled from left to right as new characters are sent.

As I began to test my sketch, I discovered that it was not sending the correct CW for certain characters. Testing each character individually, I discovered that the `Morse` library was not converting these characters to CW correctly. Digging through the library, I found that the CW translation for the characters in question was incorrect. I'm not the world's best at deciphering C++ code, but once you get used to it, you can usually get yourself in the area you're looking for, so there's really no reason to fear libraries if you have to go digging into them for any reason. After figuring out how the library converted the text to CW and correcting the encoding value for the incorrect characters in the `Morse.h` file, the library would now send the correct CW for the desired characters. When you build this project, be sure to use the `Morse.h` library in **Appendix A** or download this library from **www.w5obm.us/Arduino**.

In the Setup portion of the sketch, we need to populate our `Code[]` array with the letters, numbers and symbols to send. Rather than load each character individually, a `for()` loop is used to cycle through the ASCII values for the characters to place into the `Code[]` array. On the Arduino, the pointer to the first element in an array starts at 0, ie `Code[0]`, and increments upward from there. Since the numbers are decimal ASCII values 48-57 and the symbols for comma, dash, period, and slash are 44-47, we can have the first `for()` loop place the values 44-57 in the first part of the `Code[]` array. A second `for()` loop places the ASCII values of 65-90 for the characters A-Z in the next part of the `Code[]` array, and finally, the ASCII value of 63 for the question mark is placed as the final element in the `Code[]` array. The `Code[]` array now contains 41 entries, each corresponding to a different ASCII character value.

```
// populate the array containing the characters to use (0-9 , . / ? A-Z)
for (int x = 0 ; x<14 ; x++)    // 0-9,-./
{
  Code[x] = char(44 + x);
}

for (int x = 14 ; x<40 ; x++)   // A-Z
{
  Code[x] = char(51+x);
}

Code[40] = char(63);   // add ? character
```

The last step in the Setup loop is to randomize, or "seed", the Arduino random number generator we will be using to randomly select the character to send. The Arduino, like many computers, does not generate a true random number. Instead, the Arduino `random()` function returns a pseudo-random number that will repeat the exact same sequence of random numbers every time the Arduino resets. To add a more realistic randomness to the sequence, you can seed the random number generator using the `randomSeed()` function with a value, thereby altering the starting point in the pseudo-random sequence. A simple way to randomize the Arduino random number generator is to read an unconnected analog input pin and use the value returned as the seed value for the random number generator.

```
// randomize
randomSeed(analogRead(1));  // Seed the Random Number Generator
```

The main loop is where it all comes together. The first thing we do is read the value of the SPEED potentiometer. The SPEED potentiometer varies the voltage on the analog input pin from 0 to 5 V. The Arduino 10-bit analog-to-digital converter (ADC) will convert this voltage into a value from 0 to 1023. Rather than trying to figure out what voltage equates to the desired speed, we can use the `map()` function to do the work for us. For this project, I wanted the code speed to be variable between 5 and 35 words per minute (WPM).

```
// Read the potentiometer to determine code speed
key_speed = map(analogRead(0),0,1023,5,35);
```

The `map()` function translates the value of the SPEED potentiometer analog input to a number from 5 to 35. This number is used to set the speed used by the Morse library in words per minute. For this project, the speaker is connected to digital I/O pin 11, so the Morse library is configured to send a tone to the speaker connected to that pin.

```
// Set the Code Library to Beep on Pin 11 at the selected Key Speed
Morse morse(beep_pin, key_speed, 1);
```

Once we have set up the Morse library for use, we next need to randomly select the ASCII value of the character we want to send. Since we know we have 41 elements in the `Code[]` array, we can have the Arduino random number generator select a random number between 0 and 40. This value will be used as the index of the `Code[]` array and the ASCII value of the selected character to send is placed in the variable c.

```
// Randomly pick a character from the character array
index = int(random(41));
c = Code[index];   // Assign the value of the selected character to c
```

Before the character is actually sent, it is displayed on the first line of the LCD along with the sending speed displayed on the second line. As I mentioned at the beginning of the project, many of the libraries for the Hitachi HD-44780 LCDs don't handle the scrolling correctly, so I had to create my own code to scroll only the first line of the LCD from right to left. The `text` variable contains the last 16 characters that have been sent. When sending the next character, the first character in the `text` string variable is trimmed off using the

Arduino `substring()` function and the current character to send is added to the end of the `text` string. The string is then sent to the LCD display, which will then appear to be scrolling the text on the top line.

```
// Assign the text to display on line 0.
// When length = 15, trim and add so display appears to scroll left
 if (text.length() >15)
 {
   text = text.substring(1,16);  // Drop the First Character
 }

 text = text + String(char(toupper(c))); // Add the character to the string
 lcd.setCursor(0,0);   // Set the LCD cursor to 0,0
 lcd.print(text);   // Display the CW text
```

Finally, the character is sent to the `Morse` library, which will send the character in Morse code on the speaker.

```
morse.send(c);   // Send the character in CW
```

When the character is sent, the Arduino will repeat the main loop and continue to send and display random Morse code endlessly.

Finishing Touches

Once I have the project working on the breadboard, I like to move the finished project into an enclosure so it can be put to actual use. For most of my projects that use the 16-character by 2-line LCD, I like to put them in the Clear Enclosure for pcDuino/Arduino from SparkFun (part number: PRT-11797). This enclosure is designed for the Arduino Uno footprint and also has a handy Extension Plate (part number: PRT-11798) that adds an extra ⅝ inch of overall depth to the enclosure, allowing a little extra headroom for shields and other components that might not fit inside the standard enclosure.

Before I move the project from the breadboard to the enclosure, I create the final working version of the project schematic using CadSoft's free *EAGLE Light Edition* Schematic Editor. This program allows you to create professional looking schematics in a matter of minutes. Once your schematic is complete, you can also create a PC board design file that can be used to have your finished design etched by a commercial PC board etching service. The free version of *EAGLE Light Edition* limits you to noncommercial use; board size is limited to 100 by 80 mm (4 by 3.2 inches), and a maximum of two layers on a single sheet drawing. *EAGLE Light Edition* runs on *Windows* XP/Vista/*Windows* 7, Mac *OS X*, and *Linux*.

Using my printed schematic as a guide, I wire up the enclosure and the Arduino Uno. I like to use a prototyping shield (**Figure 7.5**) stacked on top of the Arduino to wire up my projects. I then solder the components and header pins to the protoshield. The header pins are used to connect to the external components that are mounted to the enclosure itself using DuPont-style 2.54 mm spaced cable housings and socket pins. This allows me to easily

Figure 7.5 — Random Code Practice Oscillator assembled on the prototyping shield.

remove the prototyping shield to correct the inevitable wiring errors and make any modifications to the project. You will notice that there are white marks on one end of the board headers and connector sockets. These marks are to indicate pin 1 on the various connectors to help keep me from connecting things up backward and letting the smoke out.

After wiring up the prototyping shield, I install it and the Arduino Uno into the enclosure, build cables (old PC floppy drive cables work great for this) and connectors for the external components, and mount the external components to the enclosure.

The clamshell style of the enclosure for this project makes troubleshooting easier, as you can have your project open to access the various pins and connections for troubleshooting while the project is powered up and running. **Figures 7.6** and **7.7** show the finished Random Code Practice Oscillator.

When you've finished your wiring and testing, button up the enclosure and you now have a fully functional Random Code Practice Oscillator in a nice-looking rugged case. One of the more interesting side effects of this project came when I was doing some final testing and letting it run on the test bench while I was soldering another project up at the soldering bench. After about an hour of hearing random CW in the background, I found myself copying the code in my head. Leaving the project running, I would turn the speed up a notch every now and then. At the end of just a few hours, my code speed had gone from about 10 WPM up to 18 WPM and I was copying it 100%.

Continuing the experiment, I let it run in the background for a day or so, gradually increasing the speed. At the end of this experiment in mental osmosis,

Figure 7.6 — Inside view of the Random Code Practice Oscillator.

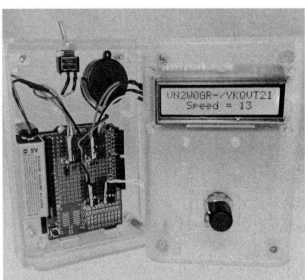

Figure 7.7 — Inside view of the Random Code Practice Oscillator showing the two-line LCD display in action.

I was copying about 23 WPM in my head. Since the characters are random, you can't guess what the next letter will be, which apparently enhances the learning process, as you have to copy each character as it was sent. Now, if I can get my sending speed to match my receiving speed, I'll be ready in time for next Field Day.

Enhancement Ideas

As we discussed in the introduction to this book, the projects are functional and complete as they are, but there will be room for you to take the existing project and make it better. As it stands, the Random Code Practice Oscillator is pretty complete. There's really not a whole lot that I would do to add features to it. One possible enhancement would be to replace the Arduino Uno with a Nano or similar version of the Arduino and shrink it all down into a smaller box and use a smaller display such as the Nokia 5110 or one of the tiny OLED displays. I'll leave the packaging up to you. Enjoy!

References

Arduino Playground — **playground.arduino.cc**
CadSoft — **www.cadsoftusa.com**
Fritzing — **www.fritzing.org**
SparkFun Electronics — **www.sparkfun.com**

CHAPTER 8

CW Beacon and Foxhunt Keyer

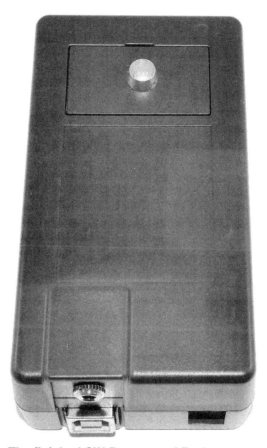

The finished CW Beacon and Foxhunt Keyer mounted in a SparkFun project enclosure.

Continuing on the CW theme, my friend and fellow Arduino builder Tim Billingsley, KD5CKP, operates a 10 meter beacon on 28.215.5 MHz. Whenever 10 meters opens up, he gets reports from all over the world with just 3 W. Unfortunately, the beacon is finicky at times and has to be kicked back to life on a regular basis. At the same time, there has been an upswing in interest among the local clubs to conduct transmitter hunts (foxhunts). This sounded like a perfect opportunity to build a small CW Beacon and Foxhunt Keyer with the Arduino.

Starting out with the usual flowchart (**Figure 8.1**), I originally designed a monster, with a scrolling organic LED (OLED) display and all sorts of bells and whistles. Soon, my design had expanded beyond the capacity of the small Arduino enclosure I had originally envisioned for it, so I had to go back to the drawing board and rethink exactly what I was trying to accomplish. In the end, with the consideration that this was to be a small, rugged, portable device, the design was simplified to be a simple beacon keyer to drive a keying relay and CW tone generator to create the modulated CW tones needed to key a 2 meter handheld transceiver.

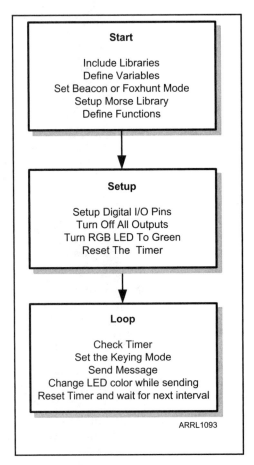

Figure 8.1 — CW Beacon and Foxhunt Keyer flowchart.

Keeping the concept as simple as possible, I next created the circuit block diagram in **Figure 8.2**. To display the status of the project, I decided to use an RGB LED (red-green-blue), allowing me to display the various states of the keying cycle. Since the time intervals, mode, and beacon text would not need to be switch-selectable, they would be hard-coded within the sketch and could easily be changed as needed.

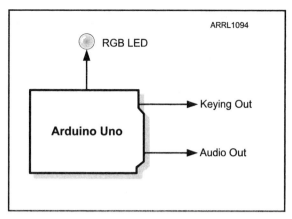

Figure 8.2 — CW Beacon and Foxhunt Keyer block diagram.

Figure 8.3 shows the Fritzing diagram for the project. To keep the beacon keyer as versatile as possible, a signal relay driven by a 2N2222A transistor was used to isolate the Arduino keying signal from the transmitter. This circuit is a good idea anytime you want to drive a relay, as the Arduino digital I/O pins can only drive 40 mA. Using a transistor to key a relay requires much less current

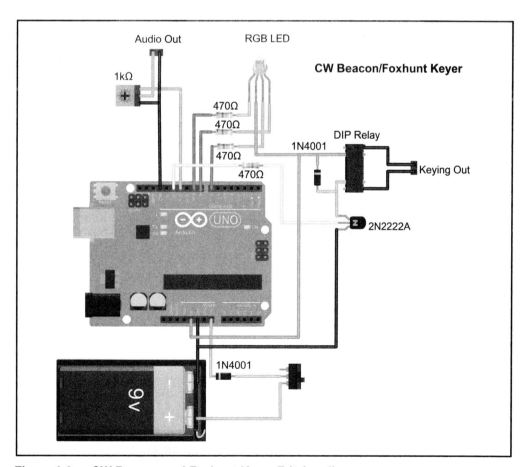

Figure 8.3 — CW Beacon and Foxhunt Keyer Fritzing diagram.

CW Beacon and Foxhunt Keyer 8-3

and helps keep you from drawing too much current and damaging your digital I/O pin. A relay also isolates the Arduino I/O pin from your transmitter, keeping potentially damaging voltages at bay. This keying relay will be used to either act as a CW key for the beacon transmitter or as the push-to-talk (PTT) keying on the foxhunt radio.

A potentiometer was placed on the audio output to allow the output level to be adjusted to whatever signal level the foxhunt transmitter would need. Note that each color section of the RGB LED has its own current-limiting resistor. At first glance, you might be tempted to use just one resistor on the common-anode side. While this works if you only have only one LED on at a time, if you ever want to start blending colors, you will need separate current-limiting resistors of different values to properly set the current for each LED as each color requires a slightly different resistor value for optimum performance and color intensity. I got in the habit of using a current-limiting resistor on each LED, just case I ever do want to blend or fade between colors at a later date.

Creating the Sketch

Once I had laid out the circuit on my breadboard to match the Fritzing diagram, it was time to write the Arduino sketch. The entire CW Beacon and Foxhunt Keyer sketch and the `Morse` library are in **Appendix A** and at **www.w5obm.us/Arduino**. The CW Beacon portion of the sketch was to repeat the beacon message at a selected interval, while the Foxhunt section was to repeat the message for a selected interval, then go silent for different time interval. Since the Arduino does not have a real time clock, there would need to be some way to keep track of time. Fortunately, for a simple project, high accuracy in the timing is not critical, so I was able to use a simple function to calculate time intervals using the `millis()` function.

As with the Random Code Practice Oscillator project, at the heart of everything all we are doing is converting text to a CW message, so we can use the same `Morse` library we used in that project. Since there is no display, the RGB LED would be used to indicate the status. Green would indicate that the keyer was operational and waiting to send a message. Blue would be used to indicate that the keyer was sending a beacon message, and red would be used to let us know that the keyer was sending the modulated CW audio foxhunt message. The LED on/off and color selection will be handled by the `ledOff()`, `ledRed()`, `ledGreen()`, and `ledBlue()` functions so that we don't have to control all three LED digital I/O lines from within the sketch itself.

Starting out with the sketch, to keep things simple and readable we assign definitions to the LED pins. This will allow you to program the digital I/O pins associated with the LED pins by color rather than by a number.

```
#define blue 7      // Blue Led Pin
#define green 8     // Green LED Pin
#define red 9       // Red Led Pin
```

Next, we define the constants that set the keyer to Beacon or Foxhunt mode, the beacon message, the size of the message array, the keying speed, the timer interval, and the number of seconds to repeat the message until the timer interval expires. We also define the digital I/O pins assigned to the keyer outputs.

```
const int mode = 0;    // Mode 0 = Beacon 1 = Foxhunt
const char beacon_call[] = "DE KW5GP Beacon";  // Message to Send
const int key_speed = 20;   // CW send speed
int interval = 10;      // Timer interval in seconds
int repeat = 1; // Repeat message time in seconds - must be a minimum of 1
const int beep_pin = 11; // Pin for CW tone
int key_pin = 12;   // Pin for PTT/Rig Key
long end_time;   // Foxhunt timing
char c;   // Character to send
bool first_pass = true;   // Flag to tell us this is the first time through
unsigned long offset = 0L;   // Timer value
Morse morse(beep_pin, key_speed, 1); // Set up the Morse Library for use
int msg_length;   // variable to hold the size of the message text array
```

The `timer()` function will return the time in seconds since the timer was reset by the `TimerReset()` function. This is how the keyer keeps track of the various time intervals. When the `TimerReset()` function is called, it doesn't actually reset anything. All it does is set the `offset` variable to the current `millis()` value, which equates to the current time in milliseconds since the Arduino was last reset. We then use this offset value to calculate our timing intervals.

```
// Timer function - returns the time in seconds since last Timer Reset
unsigned long Timer()
{
  return (millis()- offset)/1000;   // return time in seconds
}

// Timer Reset function
void TimerReset(unsigned long val = 0L)
{
  offset = millis() - val;
}
```

In the `setup()` loop, the digital I/O pins for the RGB LED and keying relay are configured and turned off, the timer reset, and finally the green LED is turned on to indicate that the keyer is ready.

```
void setup()
{
  msg_length =(sizeof(beacon_call))-1;  // Calculate size of message array
  pinMode(key_pin, OUTPUT);           // set PTT/Key pin to output
  pinMode(red, OUTPUT);    // Set up the LED pins for output
  pinMode(green, OUTPUT);
  pinMode(blue, OUTPUT);
  digitalWrite(key_pin,LOW); // Turn off the PTT/Key relay
  ledOff();   // Turn off the LED
  delay(5000);   // Wait 5 seconds
  TimerReset();   // Reset the Timer to zero
  ledGreen();   // Turn on the Green LED to indicate Beacon/Keyer ready
}   // End Setup Loop
```

In the main `loop()`, the sketch checks to see if the interval has expired, or if this is the first pass through the loop since the Arduino was reset. The sketch then determines if it is in Beacon mode or Foxhunt mode, turns on the appropriate color LED, turns on the keying relay to use as PTT if in Foxhunt mode, and sends the message. When in Beacon mode, the keying relay is used as a CW key and sends the CW directly. If the keyer is in Foxhunt mode, the sketch will add a space to the CW message and then will repeat the CW message for the duration of the repeat interval.

```
// Send if the Timer has expired or if this is the first time through
if (Timer() > interval | first_pass)
{
  first_pass = false;   // Set the first pass flag to off

  if (mode == 0)   // Set the Key mode and LED for Beacon or Foxhunt
  {
    // Set Beacon Mode
    Morse(key_pin, key_speed, 0);    // Set up to key the Relay
    ledBlue();   // Turn on Blue LED to indicate Beacon Message Transmitting
  } else {
    // Set Foxhunt Mode
    Morse(beep_pin, key_speed, 1);   // Set up to send modulated CW
    ledRed();    // Turn on Red LED to indicate Foxhunt Message Transmitting
    digitalWrite(key_pin, HIGH);   // Key the PTT
  }

// If in Foxhunt mode, repeat the message until repeat time has expired
end_time = Timer() + repeat;

  // Check to make sure repeat timer has not expired (Foxhunt Mode)
```

```
while(Timer() < end_time)
{
  // Send message in the beacon_call array one character at a time
  for (int x = 0; x < msg_length; x++)
  {
    c = beacon_call[x];  // Get the next letter to send
    morse.send(c);  // Send it in CW
  }
  if (mode == 1)  // Send a space if in Foxhunt mode to separate messages
  {
    morse.send(char(32));
  } else {
    end_time = Timer()-1;
  }
}
```

After the message is sent, the timer is reset, and the keyer will turn the green LED on again to indicate that it is ready for the next message interval.

When the sketch and prototype were completed, I created the schematic (**Figure 8.4**) to use as a guide to move the electronics to an Arduino protoshield and the finished project was mounted into a Solarbotics Arduino project enclosure, as shown in **Figure 8.5**. This enclosure is small enough to be mounted to the back of a handheld radio to be a self-contained foxhunt keyer.

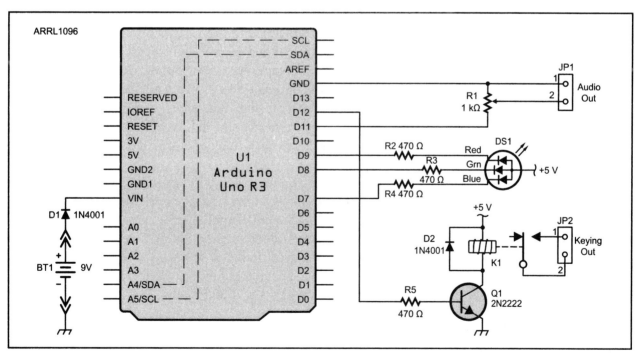

Figure 8.4 — CW Beacon and Foxhunt Keyer schematic diagram.

BT1 — 9 V battery.
D1, D2 — 1N4001 diode or equiv.
DS1 — Tri-color red, green, blue LED.
K1 — SPST or SPDT 5 V reed relay.
Q1 — 2N2222A NPN transistor or equiv.

R1 — 1 kΩ potentiometer.
R2, R3, R4, R5 — 470 Ω, ⅛ W resistor.
U1 — Arduino Uno.
Solarbotics Arduino project enclosure

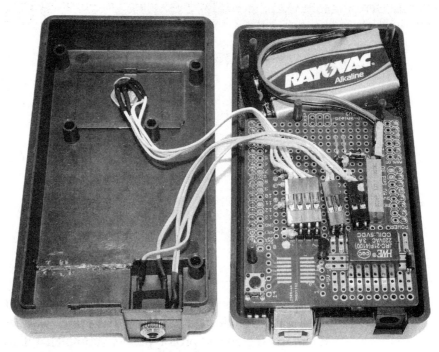

Figure 8.5 — Inside view of the CW Beacon and Foxhunt Keyer.

Enhancement Ideas

There are a number of enhancements you can add to this project. Originally, as part of the foxhunt keyer, I had planned to add a pushbutton that could be pressed to indicate that the transmitter had been found and the beacon message would be changed to a different message. Also, you can use an Arduino Nano instead of the Uno and shrink the project down to a much smaller size. You could also use a small OLED display to scroll the actual beacon message as it is being sent and to indicate when the "found" button was pressed. I would also recommend using a reed relay instead of a small signal relay to make the beacon keyer quieter.

The last enhancement I would do is probably the main reason they don't let me be the fox in foxhunts anymore. Now, since the Arduino is pretty good at robotics too, mount the handheld and foxhunt keyer on top of an Arduino quadruped robot and have it move 50 feet in random directions in between transmissions, install proximity sensors to avoid obstacles, and have motion detectors to sense when someone is nearby and have it move away from them or hunker down and go silent. Now that's a real foxhunt.

References

Arduino — **www.arduino.cc**
RadioShack — **www.radioshack.com**
Solarbotics — **www.solarbotics.com**

CHAPTER 9

Fan Speed Controller

The finished Fan Speed Controller

While fans may be a necessity with radios and other devices that generate heat, your typical fan has two speeds, on or off. Noisy fans can be distracting, especially if you don't need them running at full speed all the time. You could use a variable speed controller and control the speed of the fan, but what happens when things get too hot and you forget to turn up the fan speed?

A common method used in basic Arduino projects to control the brightness of an LED has been to use a digital I/O pin with pulse width modulation (PWM) to control the brightness by varying the duty cycle of the power to the LED. Using the same concept, we can use the Arduino to control the speed of a fan. If we add in a simple temperature sensor, we can vary the speed of the fan according to the temperature and also sound an alarm if things get too hot. This would have come in real handy two Field Days ago, when they turned me loose on 20 meters right at the same time the sun was hitting the back of the rig. It wasn't too long before the rig started to shut down from overheating and we had to set up a pair of room fans and a shade to cool the heat sink on the back of the rig to survive until the sun went down. It sure would have been nice to have a fan speed control with a heat alarm then. I'm quite sure the rig would have appreciated it. Fortunately the radio survived the ordeal, and this project is a perfect way to avoid the situation altogether the next time around.

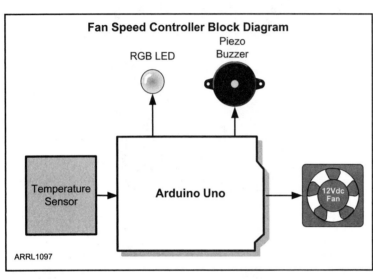

Figure 9.1 — Fan Speed Controller block diagram.

Figure 9.2 — Maxim Integrated DS18B20 temperature sensor. (The device in the photo was made by Dallas Semiconductor, now a part of Maxim Integrated.)

Starting out with our block diagram (**Figure 9.1**), you can see that we'll need a temperature sensor, a Piezo buzzer, an LED, and some method to drive the fan itself. Since the Arduino can only supply 40 mA of current per I/O pin, we will need to use a transistor between the Arduino digital I/O pin and the dc fan. An RGB LED (red-green-blue) was chosen to allow the controller to show the current fan mode and temperature range.

I feel that it is always best to prototype your circuit before writing your sketch. As you build and test the electronics, you may find something that you overlooked in your design. If you had started out with your sketch instead of the circuit, you could very well end up having to totally rewrite it if your design doesn't work as intended. In the case of the Fan Speed Controller, the dc fan wouldn't start at very slow speeds due to the PWM square waves being applied to it. A capacitor was added to smooth out the square wave a bit, and now the fan is able to start and run quietly at much lower speeds. A simple fix, but one that might have caused me fits had I not tested the design before writing the actual sketch. As I am building my circuits and starting out with the sketch, I often write smaller pieces of sketches to test out each individual function of the electronics to verify that everything works as intended. This way, if I do have problems while creating the final sketch, I know the problem is most likely in the sketch itself since the circuit functionality has already been tested with a simpler sketch. Often, I am able to cut and paste pieces of code from my testing

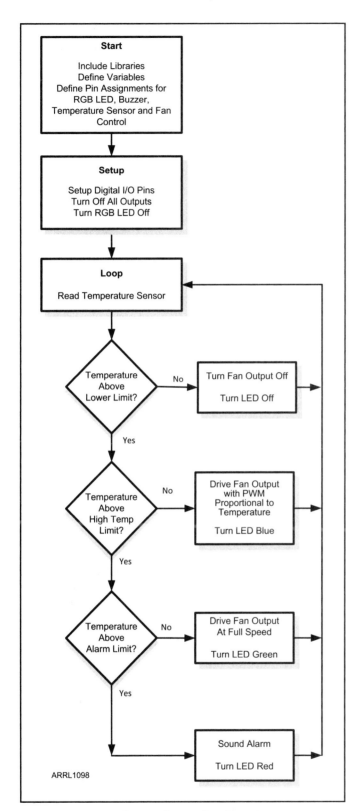

Figure 9.3 — Fan Speed Controller flowchart.

sketches into the final sketch, so I really don't lose much time in the development process.

Before we get into the actual circuit used in the fan speed controller, let's take a look at the Maxim DS18B20 temperature sensor (**Figure 9.2**). The DS18B20 is a single-component digital thermometer with a programmable resolution of either 9 or 12 bits. It has the capability for an alarm function with nonvolatile user-programmable upper and lower trigger points. Since the alarm points are non-volatile, you can set them once and they will be retained even when the sensor loses power. The DS18B20 is accurate to ±0.5 °C over the range of –10 °C to +85 °C (14 °F to 185 °F).

The DS18B20 communicates with the Arduino using the Maxim 1-Wire interface and can be powered parasitically, allowing you to use just two wires to connect it to your Arduino. In parasitic power mode, the DS18B20 uses the signals on the data line to charge a small internal capacitor to provide the power needed to operate the sensor itself. Each DS18B20 has a unique 64-bit serial number embedded in the chip, allowing you to have multiple DS18B20s on a single interface. The 1-Wire interface operates reliably at a distance of over 100 meters, allowing you to read multiple temperature sensors over a wide area with just a single digital I/O pin on your Arduino.

For the fan speed controller, the plan was for the Arduino to read the temperature sensor and set the fan speed according to the temperature that was read. The fan controller would have four basic modes of operation as shown in the flowchart for the project (**Figure 9.3**). These modes would be off, PWM fan speed proportional to the temperature, full speed, and alarm. For test purposes, 80 °F was chosen as my lower temperature limit, 100 °F as the upper temperature limit and the alarm would

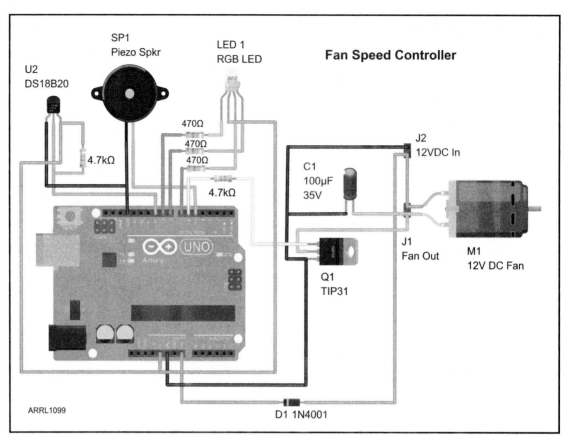

Figure 9.4 — Fan Speed Controller Fritzing diagram.

sound when the temperature reached 120 °F. At temperatures below 80 °F, the fan and LED would be off. When temperatures get above 80 °F, the fan runs at a slow speed and gradually increases in speed as the temperature rises. In this mode, the LED glows blue. When the temperature reaches 100°F, the pulse width modulation is discontinued and the fan runs at full speed. In this mode, the LED glows green. Finally, when the temperature reaches 120 °F, the LED glows red and the alarm buzzer sounds until the temperature drops below 120 °F. Note that there is no special significance to the temperatures chosen for this sketch. These values were chosen simply for testing purposes. You should set your temperature and alarm limits to suit your cooling needs.

Figure 9.4 shows the Fritzing diagram for the fan controller prototype. I chose a 12 V dc fan since that is the most commonly available dc fan, easily scavenged from old computer power supplies, and are available in a variety of sizes and speeds. This circuit can drive a fan up to 40 V dc. Higher voltage dc fans can be used simply by replacing the TIP31 transistor with a higher voltage transistor. If you do use a voltage higher than 20 V, be sure to provide an alternate method of powering the Arduino, since this design taps off of the 12 V fan power supply to power the Arduino. In this design, I chose to power the DS18B20 temperature sensor in the standard way, using the recommended 4.7 kΩ pull-up resistor between the power and data lines of the sensor.

The Sketch

The actual sketch for this project is a relatively simple one, demonstrating the power of pre-written libraries and functions. In the top of the sketch, we include the `OneWire` library found in the Arduino Playground (**playground.arduino.cc**) and define the digital I/O pins and temperature limits used by our circuit and the `OneWire` library. The complete Fan Speed Controller sketch and the `OneWire` library can be found in **Appendix A** or downloaded from **www.w5obm.us/Arduino**.

```
#include <OneWire.h>    // Include the OneWire Library

#define temp_sensor 10  // Temperature Sensor attached to Pin 10
#define red 7   // Red LED on pin 7
#define blue 8  // Blue LED on pin 8
#define green 9 // Green LED on pin 9
#define fan 6   // Fan drive on pin 6
#define alarm 5 // alarm buzzer on pin 5
#define low_temp 80   // temperature to start fan
#define high_temp 100 //  temperature to turn fan on full
#define alarm_temp 120  // alarm temperature

int DS18S20_Pin = temp_sensor; //DS18S20 Signal pin on temp sensor pin

//Temperature chip I/O
OneWire ds(DS18S20_Pin);  // on temp sensor pin
```

In the `setup()` loop, we set the pin modes for the digital I/O pins. If you organize your Arduino digital I/O into groups of inputs and groups of outputs, you can simplify the setup process by using a `for()` loop to set the pin modes and assign their starting values.

```
for (int x = 5; x <=9; x++)   // Set the Pin Modes for Pins 5 through 9
{
  pinMode(x, OUTPUT);
  if (x <= 6)  // For pins 5 and 6 start Low, 7-9 start High (LED off)
  {
    digitalWrite(x, LOW);
  } else {
    digitalWrite(x, HIGH);
  }
}
```

In the main `loop()`, we use the `GetTemp()` function to read the current temperature from the DS18B20. This function, found in the `OneWire` library example sketches, does all the dirty work to read the temperature sensor and returns the current temperature in Celsius. Since I'm not all that great at converting Celsius to Fahrenheit in my head, I added the conversion formula to the `GetTemp()` function so that it would return the temperature in Fahrenheit instead.

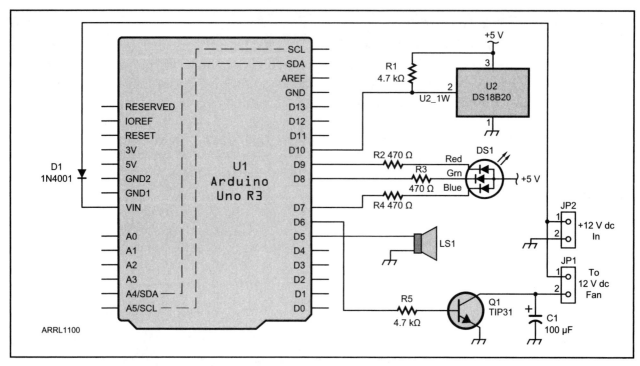

Figure 9.5 — Fan Speed Controller schematic diagram

C1 — 100 µF, 35 V electrolytic capacitor.
D1 — 1N4001 diode
DS1 — RGB common anode LED.
LS1 — Piezo buzzer.
Q1 — TIP31 NPN power transistor.

R1, R5 — 4.7 kΩ, ⅛ W resistor.
R2-R4 — 470 Ω, ⅛ W resistor.
U1 — Arduino Uno.
U2 — Maxim Integrated DS18B20 temperature sensor.

The rest of the main `loop()` uses a series of `if()` statements to set the fan operating mode and LED settings for the current temperature. I found in my testing on the breadboard that a starting PWM setting of 30 was the lowest setting that could reliably be used to start my fan at its slowest speed. Any setting slower than that and the fan would not start reliably. A `map()` statement is used to map the PWM setting from 30 (minimum) to 255 (fully on) based on the temperature range set by the upper and lower temperature settings.

```
if (temperature < low_temp)  // Everything off if below low_temp
  {
    ledOff();
    digitalWrite(fan, LOW);
    digitalWrite(alarm, LOW);
  }

// Run the Fan as a proportion of the temp
if (temperature >= low_temp && temperature <= high_temp)
{
// Starting PWM should be 30 or above to prevent fan stall
   analogWrite(fan, map(temperature,low_temp, high_temp,30,255));
   ledBlue();   // Indicate Fan Running - Temp within range
  }
```

```
// Turn the Fan on full speed
if (temperature > high_temp && temperature < alarm_temp)
{
  analogWrite(fan,255);   // Set the Fan to Max Speed
  ledGreen();   // Indicate Fan Running - Max Speed
}

// Sound the alarm
if (temperature >= alarm_temp)   // Overtemp Alarm
{
  ledRed();   // Indicate Temp over Limit
  digitalWrite(alarm,HIGH);   // Sound the Alarm
} else {
  digitalWrite(alarm, LOW);
}
```

When the breadboard prototype and sketch were debugged and complete, the schematic diagram in **Figure 9.5** was used to solder the project onto a prototyping shield, and the project was mounted inside a Solarbotics project enclosure for final testing. **Figure 9.6** shows the finished project.

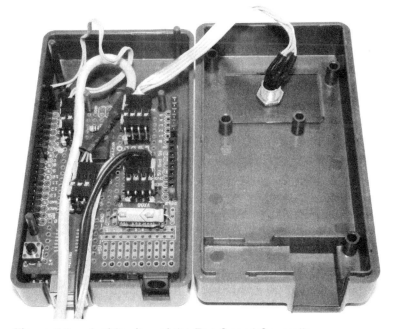

Figure 9.6 — Inside view of the Fan Speed Controller.

Enhancement Ideas

The design of the Fan Speed Controller lends itself very nicely to being mounted in a much smaller package and controlled by an Arduino Nano or similar small Arduino. Since the controller is powered by the fan's power, you can mount everything on a small perfboard attached directly to the fan. Also, as my ears would attest during testing, having an alarm mute button wouldn't be a bad idea either.

Finally, this would be an ideal project to learn about interrupts and the internal alarm settings on the DS18B20. Rather than have the sketch constantly read the temperature, you could use interrupts for your low and high PWM temperature points and have the interrupt handler manage the fan speed. One thing is for sure, I won't be cooking any more radios at Field Day as long as I have this little thing around.

References

Arduino Playground — **playground.arduino.cc**
Maxim Integrated — **www.maximintegrated.com**
Solarbotics — **www.solarbotics.com**

CHAPTER 10

Digital Compass

The finished Digital Compass mounted in a SparkFun pcDuino/Arduino enclosure.

You've got all your gear at that ideal remote site you've been planning to operate from for years. The gang has the portable tower all set up and ready to go. Just one minor detail, which way is north? How can you properly align your antennas if you don't know which way to orient it? Yeah, okay, you've probably got a compass cell phone app, handheld GPS or even a real compass, but work with me here. Wouldn't it be cool to just whip out your Arduino-powered Digital Compass to show the way?

The Honeywell HMC5883L Digital Compass (**Figure 10.1**) is an I²C-based triple-axis magnetometer combined with a 12-bit analog converter used to read magnetic field strength. The HMC5883L measures the magnetic field on three separate magneto-resistive sensors, with eight programmable gain settings and a resolution of 1 to 2 degrees. The HMC5883L is available as a module from SparkFun Electronics (**www.sparkfun.com**), DFRobot (**www.dfrobot.com**), as well as other online Arduino parts suppliers. The HMC5883L is a 3.3 V chip, so care must be taken to provide the proper voltage and signal levels to the module. Some of the more recent modules include a 5-to-3.3 V converter and the recommended I²C pull-up resistors on the module itself. This is the version of the module used in this project.

As the block diagram in **Figure 10.2** shows, once again the libraries and example sketches make this a quick and easy project to construct. A new feature in this project as compared to the previous projects is that there are now two

Figure 10.1 — The HMC5883L 3-Axis digital compass module.

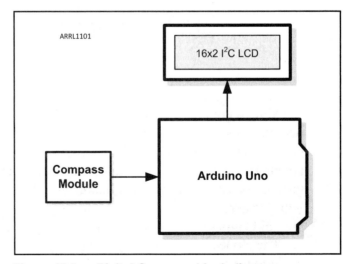

Figure 10.2 — Digital Compass block diagram

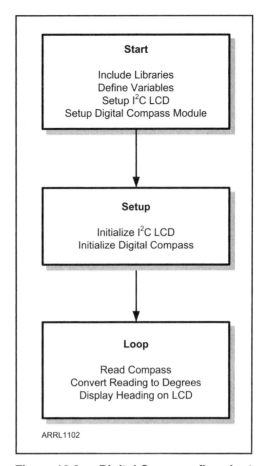

Figure 10.3 — Digital Compass flowchart

devices sharing the I²C bus, the HMC5883L compass module and the LCD module, demonstrating the versatility of the I²C bus. With both of the external components of this project sharing the I²C bus, only 8 wires (including power and ground) are required to build the project.

The `HMC5883L` and `LiquidCrystal_I2C` libraries also simplify the flowchart (**Figure 10.3**). After the initial setup and initializing the modules, it is simply a matter of using the library functions and very little in the way of actual coding is needed. In fact, the logic used to determine the 16 compass points (N, NNE, NE, and so on) and display the compass bearing on the LCD takes up more code than the rest of the entire sketch.

As mentioned above, only eight wires, plus the battery connections, are all that is needed to construct this project. **Figure 10.4** shows the Fritzing diagram of the completed project as it was built on the breadboard to begin coding. You will notice that this project appears to be missing the recommended 4.7 kΩ pull-up resistors on the I²C bus. The HMC5883L module has the I²C bus pull-up resistors on the module itself, so there is no need for the usual external pull-up resistors.

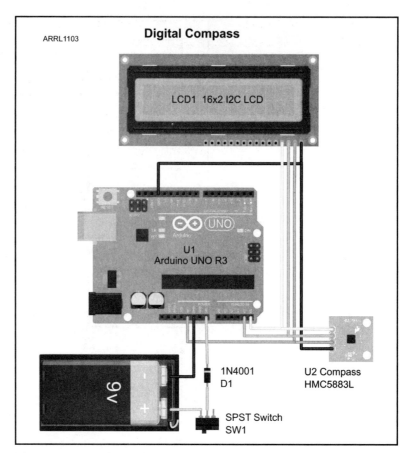

Figure 10.4 — Digital Compass Fritzing diagram

Creating the Sketch

The complete Digital Compass sketch and libraries can be found in **Appendix A** or downloaded from **www.w5obm.us/Arduino**. Using the HMC5883L library example code as a guide, the only real items of note in the initialization and `setup()` loop of the Digital Compass sketch pertain to setting the scale (gain) and mode of the HMC5883L chip. The unit used to measure the strength of a magnetic field is known as the Gauss, named after the German mathematician and physicist Carl Friedrich Gauss. For this sketch the range of the compass module is set to ±1.3 Gauss (G). For reference, the Earth's magnetic field measures between 0.31 and 0.58 Gauss at the surface.

```
// Set the scale of the compass to +/- 1.3Ga.
error = compass.SetScale(1.3);

// Set the measurement mode to Continuous
error = compass.SetMeasurementMode(Measurement_Continuous);
```

The main `loop()` reads the scaled Gauss values from Digital Compass module and stores the results in the `scaled.XAxis`, `scaled.YAxis`, and `scaled.ZAxis` function variables. Since Arduino functions can only return one value at a time, the `MagnetometerScaled Datatype` is defined in the library to pass the three `Axis` values back to the main loop.

```
// Retrieve the scaled values (scaled to the configured compass scale).
MagnetometerScaled scaled = compass.ReadScaledAxis();
```

The actual compass bearing is calculated using the arctangent function, `atan2`. This function is used to convert two numbers to their polar coordinates, which will give us the result in degrees. These results are stored in the `heading` variable. Fortunately for us, the `atan2` function handles this for us, so we don't have to remember all that trigonometry we forgot years ago.

```
float heading = atan2(scaled.YAxis, scaled.XAxis);
```

Finally, once we have calculated our compass bearing, we have to add in the magnetic declination that is specific to our location. Magnetic declination, also known as magnetic variation, is the angular different between the geographic North Pole, also known as True North, and the magnetic North Pole. This deviation value needs to be applied to the magnetic compass heading to calculate True North. You can find the magnetic declination value for your location at **www.magnetic-declination.com**.

```
float declinationAngle = 0.0169296937;   // Declination for Southaven, MS

heading += declinationAngle;
```

Next, we correct the heading if it is a negative value or if the value goes beyond 360°.

```
if(heading < 0)   // Correct for when signs are reversed.
{
  heading += 2*PI;
}

if(heading > 2*PI)   // Check for wrap due to addition of declination.
{
  heading -= 2*PI;
}
```

Finally, we convert the heading value from radians to degrees and display the value on the LCD.

```
// Convert radians to degrees
float headingDegrees = heading * 180/M_PI;    for readability.

// Output the data to the LCD display
Output(raw, scaled, heading, headingDegrees);
```

The `Output()` function displays the heading and direction on the LCD. The bearing is broken down into 16 directions, (N, NE, ENE, etc.) and the display is updated once per second.

Digital Compass 10-5

```
// Output the data to the LCD
void Output(MagnetometerRaw raw, MagnetometerScaled scaled, float heading, float
headingDegrees)
{
  lcd.clear();
  lcd.print("^   ");  // Print an up arrow to indicate compass pointer
  lcd.print(headingDegrees,1);  // Display the Heading in degrees
  lcd.print(" Deg");
  lcd.setCursor(15,0);
  lcd.print("^");

  // Calculate Direction
  Direction = " ";
  lcd.setCursor(6,1);

  if (headingDegrees >= 348.75 | headingDegrees <11.25)   // Direction = North
  {
    Direction = " N";
  }
  if (headingDegrees >= 11.25 && headingDegrees <33.75)   // Direction = North
North East
  {
    Direction = "NNE";
  }
```

Construction Notes

Once the Digital Compass was working on the breadboard, the finished schematic (**Figure 10.5**) was created and used to solder the finished project onto a protoshield (**Figure 10.6**). When you look at the HCM5883L module, you will see a small circular symbol with X and Y arrows on the board. These arrows represent the orientation of the magnetic sensors in the HCM5883L chip. The X arrow on the module should be aligned to point to your desired heading.

The completed project was mounted in a clear enclosure for pcDuino/Arduino from SparkFun (Part number: PRT-11797) along with the Extension Plate (Part number: PRT-11798). **Figure 10.7** shows the finished project.

Enhancement Ideas

One thing I discovered while constructing this project is that the HCM5883L Digital Compass is not tilt-compensated. This means that to get accurate compass readings, the compass module must be held flat, otherwise the readings may not be accurate. This can be corrected with the use of an accelerometer module and calculations can be added to compensate for compass tilt. There are several libraries and forum topics on the Arduino Playground (**playground.arduino.cc**) that add this functionality to the Digital Compass.

The simplicity of the Digital Compass circuitry lends itself easily to an Arduino Mini makeover. Using an Arduino Mini and an organic LED (OLED)

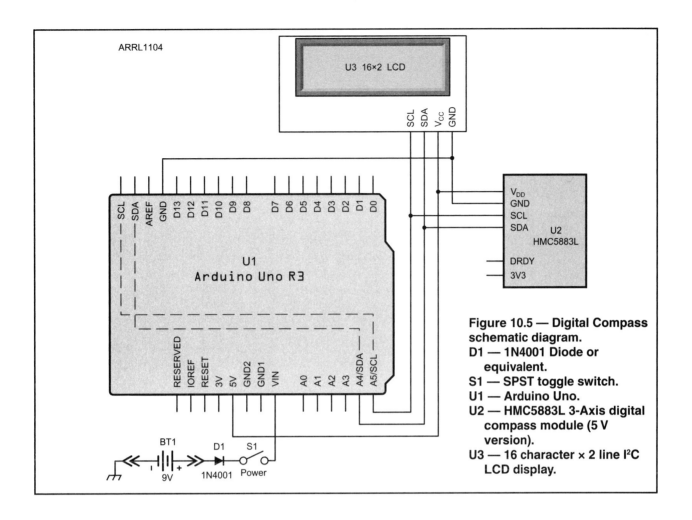

Figure 10.5 — Digital Compass schematic diagram.
D1 — 1N4001 Diode or equivalent.
S1 — SPST toggle switch.
U1 — Arduino Uno.
U2 — HMC5883L 3-Axis digital compass module (5 V version).
U3 — 16 character × 2 line I²C LCD display.

Figure 10.6 — The Digital Compass finished protoshield.

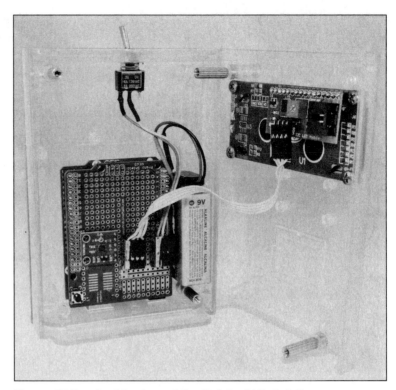

Figure 10.7 — Inside view of the finished Digital Compass.

display, you could miniaturize the digital compass to fit in a much smaller enclosure, possibly along the lines of the Altoids mint tin or even smaller.

References

Arduino Playground — **playground.arduino.cc**
bildr.blog — **bildr.org/2012/02/hmc5883l-arduino/**
DFRobot — **www.dfrobot.com**
Honeywell — **www.honeywell.com**
Magnetic-Declination — **magnetic-declination.com**
SparkFun Electronics — **www.sparkfun.com**

CHAPTER 11

Weather Station

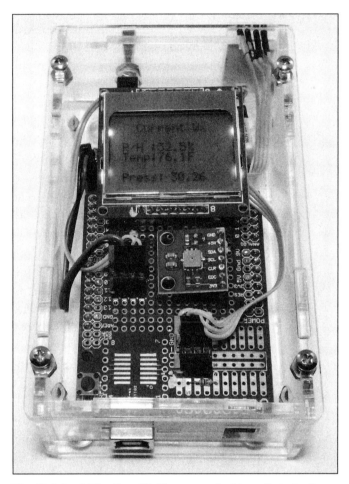

The finished Weather Station mounted in a Solarbotics Arduino Mega S.A.F.E.

As I sit here putting this all together, my weather radio is sounding all sorts of weather alerts. Almost constant severe thunderstorm, tornado watch, tornado warnings, and flash flood warnings have been going off for most of the day and the local ham emergency nets are all active. Perfect timing for a Weather Station project, wouldn't you say? Hams and weather seem to go hand in hand, from weather emergency groups and nets to plain old weather watching.

The Arduino has a wide array of sensors that can be used to monitor many different weather and atmospheric conditions. When I was growing up, our family had one of the analog wall-mounted weather stations that displayed the temperature, relative humidity, and barometric pressure on analog dials. For this project, I decided to revisit my childhood memories and give that wonderful old weather station an Arduino makeover.

Sensors

Moving forward in our Arduino adventures, this project uses a blend of two Arduino bus technologies. The MaxDetect RHT03 temperature and relative humidity sensor (**Figure 11.1**) uses the MaxDetect 1-Wire bus (not to be confused with the Maxim 1-Wire bus) while the Bosch BMP085 barometric pressure sensor uses the I²C bus to communicate with the Arduino. This project also introduces the Nokia 5110 graphic LCD display module, which is rapidly becoming the display of choice for many Arduino projects. All three devices are well supported with Arduino libraries and example code.

The MaxDetect RHT03, also available as the Aosong DHT22 and Aosong AM2302, is a pre-calibrated temperature-compensated relative humidity and temperature sensor. The relative humidity portion of the RHT03 has a resolution of 0.1% humidity and an accuracy of ±2% over a temperature range of –40 to 80 °C (–40 to 176 °F), a much better lower range than that of the Maxim DS18B20 temperature sensor used in the Fan Speed Controller. The temperature sensor in the RHT03 has a resolution of 0.1 °C with an accuracy of ±0.5 °C over the same operating range. The RHT03 communicates with the Arduino using the MaxDetect 1-Wire interface and can be placed up to 100 meters away from the Arduino for remote measurements.

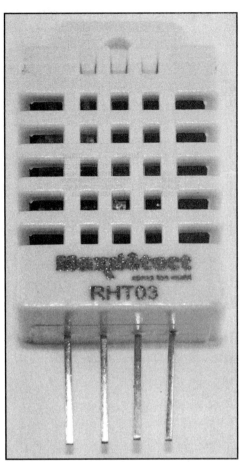

Figure 11.1 — The MaxDetect RHT03 relative humidity and temperature module.

The MaxDetect 1-Wire bus uses a different technology and is not compatible with the Maxim 1-Wire bus used by the Maxim DS18B20 and other Maxim 1-Wire devices. The MaxDetect 1-Wire interface is not really a bus, as only one sensor can be connected per Arduino digital I/O pin whereas the Maxim 1-Wire bus allows multiple Maxim devices to be connected to the same Arduino digital I/O pin. As long as you remember

Pascals, Millibars and Inches of Mercury

When I began working with the Weather Station project, I quickly discovered that the Bosch BMP085 barometric pressure sensor outputs its data in hectoPascals. Hecto what? Maybe I've led a sheltered life, but up to this point I have always known barometric pressure to be in millibars or inches of mercury. Some quick research on the web provided the formulas and information needed to convert these Pascal things to numbers I was used to.

A Pascal (Pa) is equal to 100 millibars, or, using the hectoPascals (hPa) output from the BMP085, 1 hPa = 1 millibar of pressure. One inch of mercury (inHg) is equal to 3386.489 Pascals (33864.89 hPa) at 0 °C. For reference, one standard atmosphere at sea level is defined as 29.92 inches of mercury, or 1013.25 hPa.

Since the standard pressure at sea-level can be used as a reference point, we can apply the formula in the BMP085 datasheet to our barometric pressure sensor data and calculate our altitude above sea-level. So, not only can you use your barometric sensor to measure air pressure, with just a twist you can turn it into an altimeter as well. More information may be found at **www.arduino.cc** and **www.bosch.com**.

Figure 11.2 — The Bosch BMP085 barometric pressure and temperature module.

Figure 11.3 — Weather Station output displayed on the Nokia 5110 LCD display.

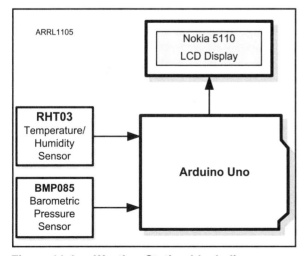

Figure 11.4 — Weather Station block diagram.

not to mix and match your MaxDetect devices and Maxim devices on the same digital I/O pin, you'll be fine.

The Bosch BMP085 barometric pressure sensor (**Figure 11.2**) communicates with the Arduino using the I²C bus. The BMP085 has resolution of 0.01 hPa (hectoPascals) with a typical accuracy of ±1 hPa over an operating temperature range of –40 to 85 °C. (See the sidebar, "Pascals, Millibars, Inches of Mercury.") As with the RHT03, the BMP085 also has an embedded temperature sensor that has a resolution of 0.1 °C with an accuracy of ±1 °C over the same operating range as the barometric pressure sensor portion of the module. The BMP085 sensor chip requires 3.3 V to operate, but many of the more recent modules have onboard 5 V regulators. This project uses the 5 V version of the BMP085 module.

The Nokia 5110 graphic LCD (**Figure 11.3**) was originally intended for use as a cell phone display. Using the PCD8544 display controller, the Nokia 5110 has an 84 × 48 pixel display, which allows up to six lines of 15 characters of text. The Nokia 5110 is a graphic LCD, meaning you can display images, charts, and graphs as well as text. It uses four small LEDs for backlighting and is currently available with either a white or blue backlight. The datasheet recommends powering the Nokia with 3.3 V; however it can be used at up to 5 V. The Nokia 5110 connects to the Arduino using five digital I/O pins, and the contrast of the display can be adjusted via software.

Starting out with the usual block diagram (**Figure 11.4**), I wanted the Arduino to read the RHT03 and BMP085 sensors and display the current temperature, relative humidity, and barometric pressure on the Nokia 5110 LCD. Since both sensors have internal calibration data that has to be applied to the raw sensor data, the Arduino would also be used to apply this calibration data to the raw sensor readings to calculate the correct results.

Using the block diagram as a guide, the next order of business was to wire the project up on the breadboard. Using the Fritzing diagram I created for this project (**Figure 11.5**), I wired up the project on the breadboard. You will notice that

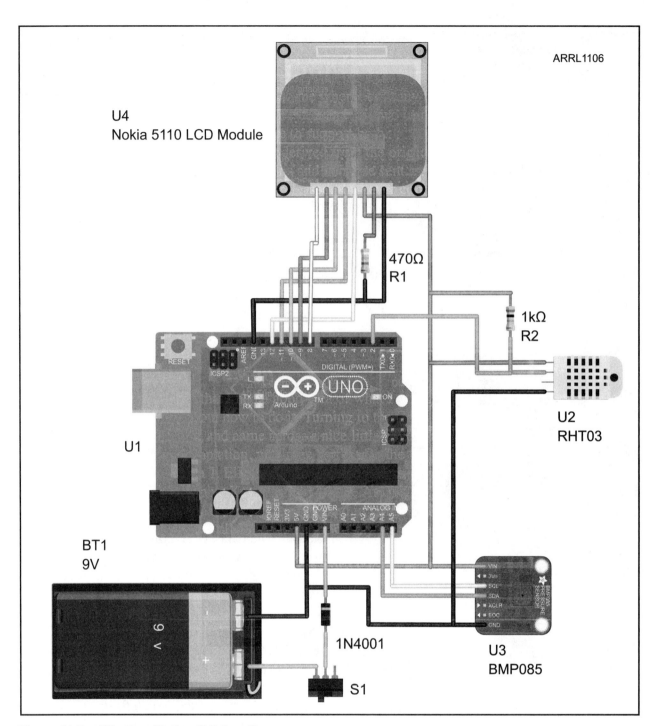

Figure 11.5 — Weather Station Fritzing diagram.

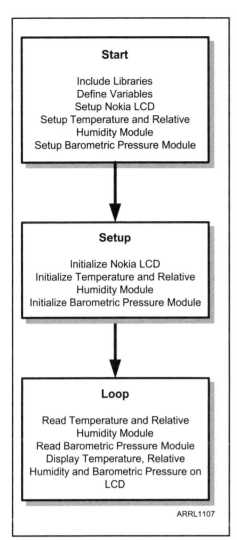

Figure 11.6 — Weather Station flowchart.

there are no pull-up resistors on the I²C bus for the BMP085. The BMP085 module used in this project is the 5 V version and comes with the I²C bus pull-up resistors mounted on the module itself. You will also note that the datasheet calls for a 1 kΩ pull-up resistor on the RHT03 data line connected to the Arduino digital I/O pin. Finally, a 470 Ω resistor is used on the backlight pin of the Nokia 5110 LCD.

Planning out the sketch flowchart (**Figure 11.6**), I found a number of libraries available for the DHT03 and the Nokia 5110 LCD. The libraries I chose to use were the DHTLib library from the Arduino Playground (**playground.arduino.cc**) and the LCD5110_Basic library by Henning Karlsen (**www.henningkarlsen.com/electronics/**) although any of the other libraries would do just as well. At the time this project was created, I could not locate a suitable library for the BMP085 sensor, so the code for the BMP085 does not use a library and the SparkFun example sketches were used as a guideline instead. Most of the complex operations are handled by the libraries or by the functions in the example sketches, which greatly simplifies the use of these sensors in your projects.

Unlike the previous projects, there is a lot of initialization needed to get the sensors and display ready for operation. In addition to the standard library #include statements, the addresses and Arduino I/O pins must be defined for use by their associated libraries. Because we are using I²C for the BMP085, we also must include the Arduino internal I²C Wire.h library. The complete sketch and required libraries for this project can be found in **Appendix A** and at **www.w5obm.us/Arduino**.

```
#include <Wire.h>   // Use the internal I2C Library
#include <dht.h>    // Use the DHT Relative Humidity Library
#include <LCD5110_Basic.h>    // Use the Nokia 5110 LCD Library
#define BMP085_ADDRESS 0x77   // I2C address of BMP085

dht DHT;   // Define the DHT object
#define DHT22_PIN 2   // Set the I/O pin used for the RHT03 Sensor

// The Nokia LCD module is connected to the following pins.
//      CLK  - Pin 12
//      DIN  - Pin 11
//      DC   - Pin 10
//      RST  - Pin 8
//      CE   - Pin 9

LCD5110 glcd(12,11,10,8,9);   // Set the I/O pins used by the Nokia display
extern uint8_t SmallFont[];   // Define the Small Font for the Nokia display
```

Next, the variables for the BMP085 calibration values need to be defined as well as the variables for the BMP085 and RHT03 raw sensor data values.

```
// Calibration values for the BMP085
int ac1, ac2, ac3, b1, b2, mb, mc, md;
unsigned int ac4, ac5, ac6;

// b5 is calculated in bmp085GetTemperature(), this variable is also used in
// bmp085GetPressure() so Temperature() must be called before Pressure()

long b5;
int chk ;    // Status Check variable for RHT03

// These variables contain the calculated results
float centigrade, fahrenheit, inHg;
```

In the setup() loop, the I²C bus and the Nokia 5110 LCD are started. A startup message is briefly displayed on the LCD display before continuing on with the rest of the setup() loop.

```
Wire.begin();    // Start the I2C Interface

glcd.InitLCD(65); // Initialize the Nokia 5110 Display, set Contrast to 65
glcd.setFont(SmallFont);    // Set the Font to Small Font
glcd.print("KW5GP", CENTER, 0);   // Display the Startup screen
glcd.print("Weather", CENTER, 8);
glcd.print("Station", CENTER,16);
glcd.print("Initializing", CENTER,32);

delay(3000);

glcd.clrScr();   // Clear the LCD screen
```

The last operation performed by the setup() loop is to run the BMP085 calibration function:

```
// Run the BMP085 Calibration Function
bmp085Calibration();
```

In the main loop(), the LCD is set up with a display template and the RHT03 is read. In case of a read error, the error is displayed on the LCD and the sensor data is ignored. This portion of the sketch introduces the Arduino switch...case() statement. The switch...case() statement allows the selection of different blocks of code to be executed based on the value of a variable, in this case, the chk variable is used to determine which branch of the code is executed. The end of each branch of the code is signified by the break keyword and program execution continues at the end of the switch...case() block of code. If no conditions match within the switch...case() statement, the `default` block of code is executed.

While both sensors are capable of providing temperature data, for this sketch, the temperature readings from the RHT03 are used.

```
glcd.print("Current Wx", CENTER, 0);
glcd.print("R/H : ",0,16);
glcd.print("Temp: ",0,24);

// Read the RHT03 RH/Temp Sensor
chk = DHT.read22(DHT22_PIN);    // Read the RHT03 RH/Temp Sensor

switch (chk)
{
  case DHTLIB_OK:
    // Display the RH Data if it's a valid read
    glcd.printNumF(DHT.humidity,1,30,16);
    glcd.print("%     ",55,16);
    centigrade = DHT.temperature;
    fahrenheit = (centigrade * 1.8) + 32;   // convert to Fahrenheit
    glcd.printNumF(fahrenheit,1,30,24);

    glcd.print("F",55,24);
    break;

  case DHTLIB_ERROR_CHECKSUM:
    glcd.print("CK Error",25,16);
    break;

  case DHTLIB_ERROR_TIMEOUT:
    glcd.print("T/O Error",25,16);
    break;

  default:
    glcd.print("Unk Error",25,16);
    break;
}
```

Next, the barometric pressure data is read from the BMP085. The BMP085 outputs the barometric pressure in Pascals. The sketch converts this into the more commonly known inches of mercury. One interesting feature of the BMP085 is that it can also calculate altitude based on the difference between standard sea-level atmospheric pressure and the barometric pressure read by the BMP085.

```
// bmp085GetTemperature() MUST be called first
float temperature = bmp085GetTemperature(bmp085ReadUT());
float pressure = bmp085GetPressure(bmp085ReadUP());
float atm = pressure / 101325; // "standard atmosphere"
```

```
float altitude = calcAltitude(pressure); //Uncompensated altitude - in Meters
pressure = pressure / 1000;   // Convert to KiloPascals
inHg = pressure * 0.2952998016471232;   // Convert KPa to Inches of Mercury
glcd.printNumF(inHg,2,40,40);   // Display the Pressure in Inches of Mercury
```

The real magic for converting the raw barometric pressure data from the BMP085 happens in the bmp085GetPressure() function. The BMP085 has eleven 16-bit calibration values stored in the BMP085's internal EEPROM memory. These calibration values are applied using a complex formula to calculate the actual barometric pressure read by the BMP085. I won't even attempt to try to understand what this function does other than state that it uses the formula provided by Bosch in the BMP085 datasheet to convert the raw data into actual barometric pressure. This is part of the fun of the Arduino. You don't have to understand how a pre-existing function or library works as long as it provides you the information you need for a given project.

```
// This function calculates pressure
// calibration values must be known
// b5 is also required so bmp085GetTemperature() must be called first.
// Value returned will be pressure in units of Pa.

long bmp085GetPressure(unsigned long up)
{
  long x1, x2, x3, b3, b6, p;
  unsigned long b4, b7;
  b6 = b5 - 4000;

  // Calculate B3
  x1 = (b2 * (b6 * b6)>>12)>>11;
  x2 = (ac2 * b6)>>11;
  x3 = x1 + x2;
  b3 = (((((long)ac1)*4 + x3)<<OSS) + 2)>>2;

  // Calculate B4
  x1 = (ac3 * b6)>>13;
  x2 = (b1 * ((b6 * b6)>>12))>>16;
  x3 = ((x1 + x2) + 2)>>2;
  b4 = (ac4 * (unsigned long)(x3 + 32768))>>15;
  b7 = ((unsigned long)(up - b3) * (50000>>OSS));

  if (b7 < 0x80000000)
    p = (b7<<1)/b4;
  else
    p = (b7/b4)<<1;

  x1 = (p>>8) * (p>>8);
  x1 = (x1 * 3038)>>16;
  x2 = (-7357 * p)>>16;
```

```
p += (x1 + x2 + 3791)>>4;
long temp = p;
return temp;
```

}

Construction Notes

Once the sketch was completed and the project working as advertised on the breadboard, the schematic diagram for the project (**Figure 11.7**) was used to

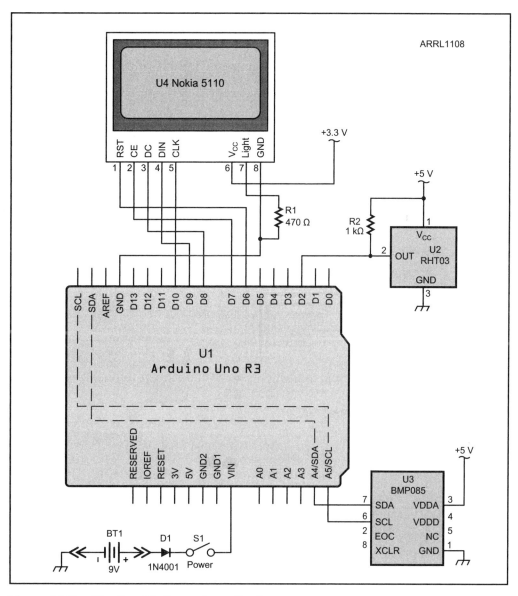

Figure 11.7 — Weather Station schematic diagram.

BT1 — 9 V battery.
D1 — 1N4001 diode.
R1 — 470 Ω, ⅛ W resistor.
R2 — 1 kΩ ⅛ W resistor.
S1 — SPST switch.
U1 — Arduino Uno.
U2 — MaxDetect RHT03 relative humidity/temperature sensor module.
U3 — Bosch BMP085 barometric pressure/temperature sensor module.
U4 — Nokia 5110 LCD display module.

build the protoshield. I mounted the finished project in a clear Solarbotics Mega S.A.F.E. enclosure. I prefer to use the Mega size enclosure as it allows more room for mounting external components such as a battery, switches, and other parts. To get a more accurate temperature and humidity reading, the RHT03 was mounted on the outside of the enclosure.

I have also found the Solarbotics S.A.F.E. enclosures to be ideal for projects that use the Nokia 5110 display, as the display can actually be mounted using a header socket onto the protoshield itself. This allows the LCD display to be clearly seen while keeping it safe inside the enclosure.

Enhancement Ideas

This is one of those projects where it can be difficult to choose what you want it to be as you have so many atmospheric and weather-related sensor options available. Since many weather-related items involve a trend over a period of time, adding a real-time clock calendar and datalogging function and graphic trends on the graphic LCD would be ideal. You can easily add a rain gauge and an anemometer for wind speed and have a nearly complete picture of the current weather. A much simpler enhancement would be to use a digital I/O pin to control the brightness of the backlight using PWM. You can keep track of the barometric pressure and add an up or down arrow next to the barometric pressure reading to indicate that the pressure is rising or falling. Finally, you could use a text-to-speech module and have the weather station speak the time, temperature, and other parameters for you.

This project turned out to be a lot of fun for me and it will come in real handy in just a few short weeks when the club does our annual exercise in insanity as we compete in the outdoor category of the annual Society for the Preservation of Ham Radio (SPAR) Winter Field Day event, where temperature is part of the contest exchange. Hopefully next year, I'll have the Ethernet side of this project worked out and we can have the contest software automatically handle the time and temperature side of things for us.

References

Adafruit Industries — **www.adafruit.com**
Aosong Electronics — **www.aosong.com**
Arduino Playground — **playground.arduino.cc**
bildr.blog — **www.bildr.org/2011/06/bmp085-arduino/**
Bosch — **www.bosch-sensortec.com**
Henning Karlsen — **www.henningkarlsen.com/electronics/**
MaxDetect — **www.humiditycn.com**
Society for the Preservation of Ham Radio — **www.spar-hams.org**
SparkFun Electronics — **www.sparkfun.com**

CHAPTER 12

RF Probe with LED Bar Graph

The finished RF Probe (right) and RF sensing unit.

One of the cool things about the Arduino is that it is a tool that can be used to make other tools. If you're like most hams, you don't often have a need for a lot of test equipment, and when you do need to test something, you can usually borrow what you need from a fellow ham. The reason for this is primarily because some test equipment can be rather expensive, or it's just not worth the investment for something you'll only use once or twice in a blue moon.

In my case, I don't often need an RF probe, but it can come in handy when you want to know if that QRP transmitter is working and you don't happen to have a wattmeter or SWR meter handy or you need to track down some stray RF in the shack. While researching another project, I came across the "RF Driven On-Air Indicator" article by Keith Austermiller, KB9STR, which itself is derived from "The 'No Fibbin' RF Field Strength Meter" by John D. Noakes, VE7NI.[1,2] With a few minor tweaks, those projects could be adapted into an RF probe that would allow an Arduino to drive an LED bar graph display instead of a meter to indicate the strength of the RF signal.

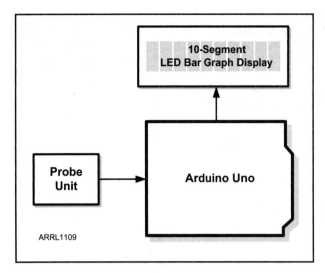

Figure 12.1 — RF Probe block diagram.

RF Sensing Unit

This is actually a two part project. First, we need to construct the RF sensing unit and then we have to connect the sensing unit to the Arduino and display unit. The RF sensing unit is built in a separate enclosure, allowing it to be adapted for other projects when not needed as an RF probe. **Figure 12.1** shows the block diagram for the project. In addition to the RF sensing input, the Arduino would be used to display the signal strength on a bar graph display with 10 LEDs. The original design called for using a Maxim MAX7219 bar graph LED driver chip. After beginning the actual circuit construction, I felt that there was no need for the added complexity of an LED driver chip and decided it would be easier to just have the Arduino drive the bar graph LED directly.

The construction of the RF Probe goes a little bit differently than the previous projects. Rather than prototype the RF sensing unit on the breadboard and have to be concerned about the effects of RF on the breadboard wiring, the RF sensing unit was to be built and mounted in a separate enclosure.

Using the schematic in the "RF Driven On-Air Indicator" article, a few modifications were made to adapt it for use in this project, as shown in **Figure 12.2**. Since the only output desired for the RF sensing unit is an analog voltage representing the signal strength, the output transistors and the relay drivers are omitted and a 1 kΩ buffering resistor added between the op amp on the sensing unit and the Arduino. Also, to improve sensitivity, the 1N34A germanium diodes called for in the circuit were replaced with 1N5711 Schottky diodes.

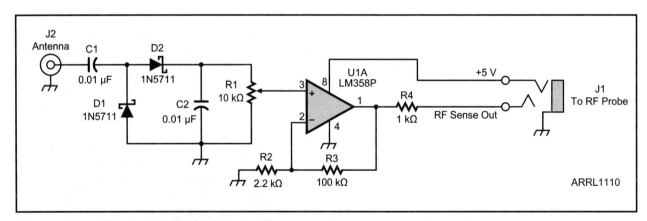

Figure 12.2 — RF sensing unit schematic diagram.
C1, C2 — 0.01 µF, 35 V capacitor.
D1, D2 — 1N5711 Schottky diode.
J1 — ⅛-inch stereo mini jack.
J2 — SO-239 chassis mount coax connector.
R1 — 10 kΩ potentiometer.

R2 — 2.2 kΩ, ¼ W resistor.
R3 — 100 kΩ, ¼ W resistor.
R4 — 1 kΩ, ¼ W resistor.
U1 — LM358P op amp.

The choice of diodes used in the RF sensing unit is critical and you will need to use either 1N34A germanium diodes or preferably 1N5711 Schottky diodes. Whatever diode you choose to use, do not use a silicon diode such as the 1N4001. A silicon diode has a forward voltage drop of approximately 0.7 V, while a germanium diode has a forward voltage drop of 0.3 V, and the Schottky diode has the lowest forward voltage drop of the bunch at 0.2 V. These small differences in forward voltage drop can significantly impact the sensitivity of the RF sensing unit.

The RF sensing unit is built on a standard protoboard cut to fit inside a Radio Shack (part number 270-1802) 4 × 2 × 1 inch project box. The antenna connector and output jack connect to the circuit board using pin headers and DuPont-style female headers allowing easy removal of the circuit board to correct the inevitable wiring error. The LM358P op amp is mounted in a socket for those times when you accidentally feed it 100 W and let the smoke out. A 10 inch piece of AWG #14 solid wire was soldered to the center conductor of a PL-259 coax connecter to serve as the RF pickup for the sensing unit. An SO-239 coax chassis connector mounted to the project box allows you to use different antennas for the RF pickup antenna. A stereo ⅛-inch mini jack is used to connect between the RF sensing unit and the Arduino assembly. **Figure 12.3** shows the finished unit.

Figure 12.3 — The inside of the RF sensing unit.

Breadboard

The next order of business is to wire up the Arduino half of the project on your breadboard using the Fritzing diagram in **Figure 12.4**. This project introduces two new I/O methods for the Arduino. We will be using the Arduino's built-in analog-to-digital converter to change the analog voltage output from the RF sensing unit into a value that is converted to a digital representation of the signal strength, and then output that to a 10 LED bar graph display. For an RF probe, you don't necessarily need the power of an LCD display, as all you are really interested in is a relative RF field strength indication — perfectly suited to the bar graph-style displays. Since the Arduino will be doing all of the work, all you really need on the protoshield is the 330 Ω current-limiting resistors for the bar graph LED display.

Creating the Sketch

The sketch for the RF Probe project is actually rather simple. Using the flowchart (**Figure 12.5**), the sketch itself is quite small, showing how well adapted the Arduino is to the simple I/O tasks used in this project. The sketch requires no libraries and is very straightforward. All the sketch has to do is read the analog voltage from the RF sensing unit and output a bar graph

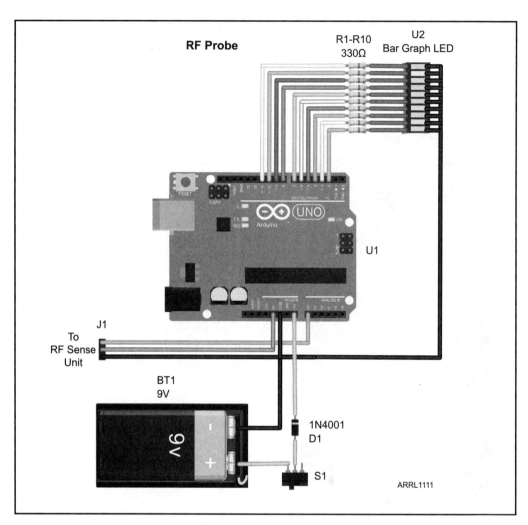

Figure 12.4— RF Probe Fritzing diagram.

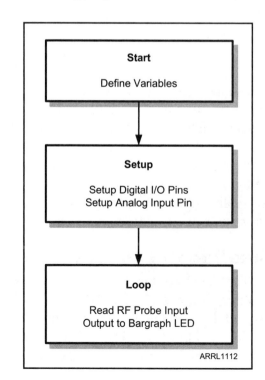

Figure 12.5 — RF Probe flowchart.

representation of the input voltage on the bar graph LED.

The bar graph LED requires 10 digital I/O pins to drive it directly. Since we're not planning for the Arduino to do anything other than drive the bar graph LED, there's no issue with using most of the Arduino's digital I/O pins in this way. Since consecutive pins were used in the design, we can use a `for()` loop to set the LED modes as outputs rather than initializing each digital I/O pin separately.

```
for (int x=2; x<=11; x++)   // Set pins 2-11 as LED Outputs
{
  pinMode(x, OUTPUT);
}
```

In the main `loop()` the analog voltage is read from the RF sensing unit and converted into a digital value between 0 and 1023, corresponding to the 0 to 5 V output of the RF sensing unit.

```
// Read the Signal Strength from the RF Sensor

  Signal_Strength = analogRead(Signal_Pin);
```

The next thing is to determine how many bars to light on the bar graph LED. Based on my RF testing, a level of approximately 3.6 V from the RF sensing unit was determined to be a good indication of maximum RF field strength. Note that with the 1 kΩ resistor between the RF sensing unit and the Arduino, 3.6 V will be the maximum value you can expect to see on the Arduino analog input pin. This level can be adjusted by varying the 10 kΩ potentiometer (R1) on the RF sensing unit.

Now would be a good time to introduce the Arduino `map()` statement. The `map()` statement is a great way to modify the scale of your data. In this case, the 0 to 5 V analog input from the RF sensing unit is mapped from a possible digital value of 0 to 1023 into a value from 1 to 10, with our maximum indication occurring at a value of 750 (approximately 3.6 V). This tells us how many LEDs on the bar graph display need to be lit to indicate the RF field strength. Using the `map()` statement in this way, the first LED in the bar graph will always be lit. This is used as a power indicator to let us know the RF Probe is powered on and ready to go.

```
// Figure out how many bars to light

// Map the Signal Strength to # of bars (1-10)

// Led one is always on to indicate power

  bars = map(Signal_Strength,0,750,1,10);
```

Finally, the mapped value from 1 to 10 is output to the bar graph LED. A `for()` loop is used to light the desired number of LEDs in the bar graph, while a second `for()` loop is used to make sure the rest of the LEDs in the display are turned off, providing a dynamic display of the RF field strength seen at the RF sensing unit.

```
for (int x = 1; x<=bars; x++)   // Turn on all bars up to the mapped bar
{
  digitalWrite(x+1, HIGH);   // Turn on the LED's
}
if (bars < 10)
{
  // Make sure the rest of the bars are off
  for (int y = bars + 1; y<=10; y++)
  {
    digitalWrite(y+1, LOW);   // Turn off the LED's
  }
}
```

Once the prototype was working on the breadboard, a protoshield with the dropping resistors and connectors for the RF sensing unit and the bar graph LED was constructed using the schematic diagram in **Figure 12.6** and everything was mounted into a SparkFun Arduino project enclosure (PRT-10088) as shown in **Figure 12.7**. The Avago Technologies HSDP-4832 bar graph LED array I used has three colors of LEDs. The first three LEDs are green, followed by four yellow LEDs and finally three red LEDs, which allows you to easily see the display values as they change with the RF field strength. Due to the simple design and sensitivity of the RF sensing unit, I have used the RF Probe to sense RF all the way up to 440 MHz.

Enhancement Ideas

This would be an excellent project for an Altoids mint tin and an Arduino Nano. If you used a BNC connector instead of the SO-239, you could possibly fit everything inside the Altoids tin with a small cutout for the bar graph LED. You could also replace the bar graph LED with a small organic LED (OLED) and calibrate the unit against a real field strength meter and have it provide a digital representation of the actual field strength. And finally, you could implement the Maxim MAX7219 LED driver to drive either a bar graph or 7 segment LED, or even display the output as a graph on an 8 × 8 LED array.

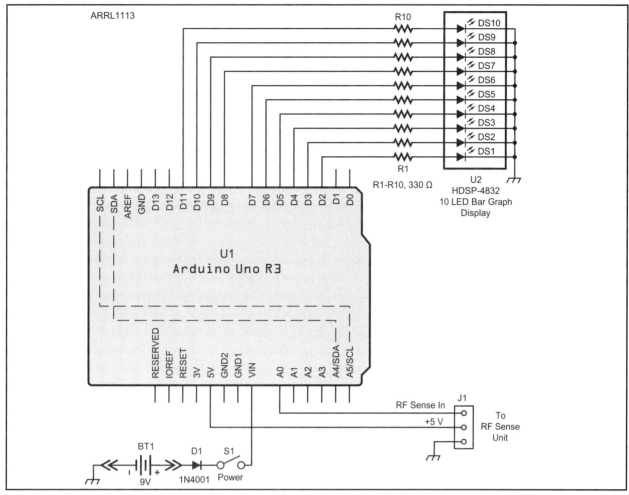

Figure 12.6 — RF Probe schematic diagram.
BT1 — 9 V battery.
D1 — 1N4001 diode.
J1 — 3-pin header with cable and plug to match RF sensing unit.
R1-R10 — 330 Ω, ⅛ W resistor.

S1 — SPST toggle switch.
U1 — Arduino Uno.
U2 — Avago Technologies HSDP-4832 10 LED bar graph array.

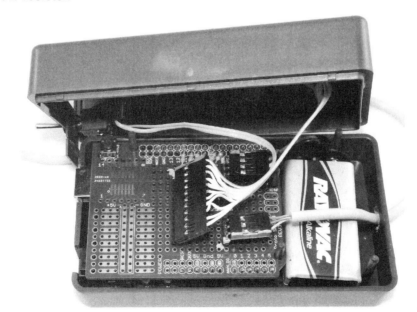

Figure 12.7 — Inside view of the RF Probe.

References

Avago Technologies — **www.avagotech.com**
Maxim Integrated — **www.maximintegrated.com**
RadioShack — **www.radioshack.com**
SparkFun Electronics — **www.sparkfun.com**

Notes

[1] K. Austermiller, KB9STR, "An RF Driven On-Air Indicator," *QST*, Aug 2004, pp 56-57.

[2] J. Noakes, VE7NI, "The 'No Fibbin' RF Field Strength Meter," *QST*, Aug 2002, pp 28-29; Feedback Sep 2002 *QST*, p 88.

CHAPTER 13

Solar Battery Charge Monitor

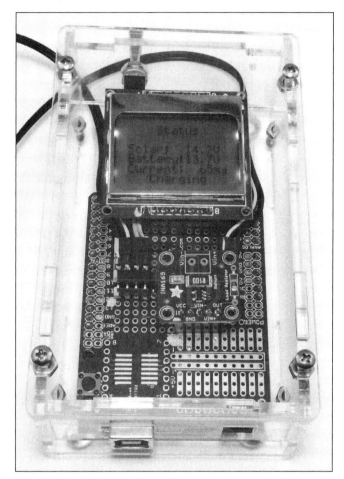

The finished Solar Charge Monitor mounted in a Solarbotics Arduino Mega S.A.F.E.

The Arduino, with its low power requirements, is a perfect candidate for solar-powered projects. The Arduino Uno has a typical current draw of approximately 55 mA, while its smaller brother, the Nano, draws a mere 20 mA. Of course, to power your solar projects, it always helps to have battery backup for those times when there's not enough sunlight to power things. The cool thing is you can have your Arduino manage the battery charging process while it's off performing other tasks. With the cost of solar panels becoming more affordable every day, you can build your own portable solar charging system for just a few dollars.

One of the major concerns when using solar cells is the wide variation of voltage and current supplied by the solar panels due to clouds or the angle of the sun on the solar panels. To maintain proper battery charge, you'll need to keep track of the voltage and current supplied to the battery from the solar panels. With most solar arrays providing 12 V or higher, trying to read this voltage directly can do horrible things to the 5 V maximum analog input on your Arduino, if not frying your Arduino itself to a crackly crunch. Fortunately, you can use a voltage divider to bring the solar array voltage and battery voltage down to a range that your Arduino can safely use.

Current Monitor

Reading the charging current between your solar array and the battery is a whole other issue. You really don't want your Arduino anywhere near the voltages that charge the battery, but you will need to read the current flowing between the array and the battery. Fortunately, someone else has faced this issue and created a module just for this purpose.

Figure 13.1 — The INA169 current sensing module.

The Texas Instruments INA169 shown in **Figure 13.1** is a high-side dc current shunt monitor that can handle up to 60 V dc. A high-side monitor can be inserted anywhere in your circuit, whereas a low-side device must be inserted between the target ground and true ground. Since the voltage drop across the shunt resistor used for measuring is proportional to the current draw, with a low-side monitor, the ground reference will change as the current varies. A moving ground reference can be a bad thing, and is not something we really want to have to take into account in our circuit designs. With a high-side current shunt monitor, we can place it anywhere in our circuit without fear. The INA169 uses a precision amplifier measuring differentially across a 0.1 Ω, 2 W precision sense resistor, allowing you to read up to 5 A continuously. The output voltage of the module that represents the current in the circuit under test is in direct proportion to the sense current. The INA169 default setting is an output of 1 volt per ampere measured, giving you an output of 0 to 5 V over the range of 0 to 5 A. The sensitivity of the current measurement can be changed simply by replacing the onboard 10 kΩ measurement resistor with a different value.

The solar panels used for this project are two 150 × 85 mm (5.9 × 3.3 inch) 7.5 V panels that output up to 220 mA. The idea was to keep this project small, light, and portable. Taking it to full-scale, I would add two additional sets of paired panels, giving it a total output of up to 660 mA, a little more than half of an amp of solar charging power from a small, portable solar array. The panels were glued to a small wooden framework and attached to a servo pan/tilt head assembly. I used a servo pan/tilt mount to allow the addition of servos and photosensors at a later date to automatically track the sun and maintain maximum charging rate as the sun moves across the sky. **Figures 13.2** and **13.3** show the solar panel assembly.

Figure 13.2 — The solar panel array mounted on the pan/tilt servo assembly.

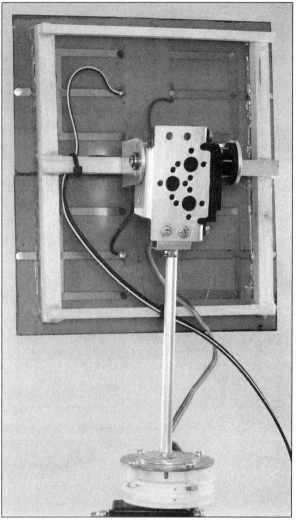

Figure 13.3 — Rear view of the solar panel array showing how it is mounted to the pan/tilt servo assembly.

Circuit Design

Now that we have a solar array that can generate our charging current, and a sensor that can read the current, we can begin designing the actual circuit. Using the block diagram (**Figure 13.4**), we'll want to have the Arduino measure the charging current and the voltage of both the solar array and the battery. With a maximum of 220 mA charging current for this project, there's really no need to regulate the charging current between the solar panels and the 5.0 A/h gel cell battery. Once you get to charging currents of 1 A or more, you will need to limit the charging current with additional external circuitry. At charging currents greater than 3 A, you will also need to beef up the blocking diode between the solar array and the battery. Since the Arduino only draws about 55 mA, we can also use the solar array to power the Arduino while having enough current left over to charge the battery.

Since we have large external components that may be carrying higher currents at some point, the solar array and battery are connected to the Arduino using

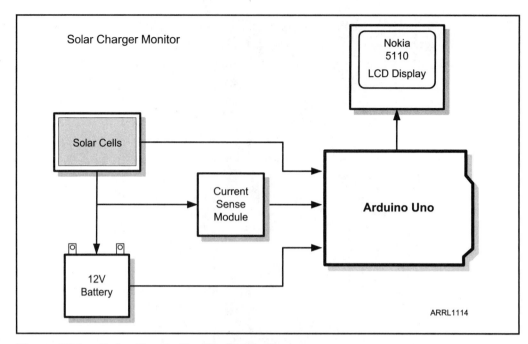

Figure 13.4 — Solar Charge Monitor block diagram.

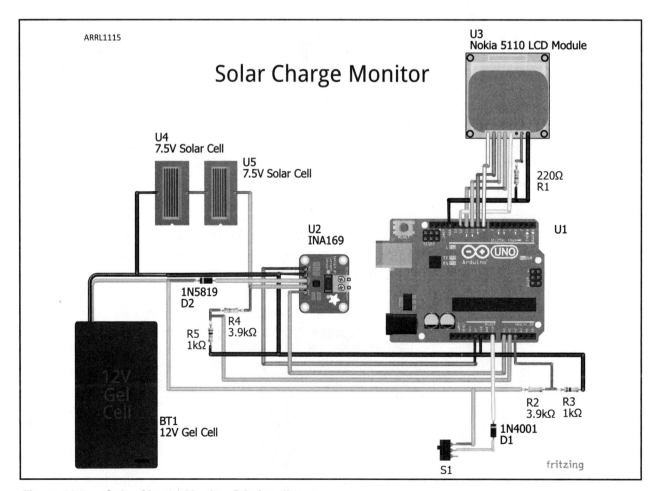

Figure 13.5 — Solar Charge Monitor Fritzing diagram.

AWG #14 wire and Anderson Powerpole connectors (see the sidebar) for quick disconnecting. If you are planning to use a larger charging current, you will want to move up to AWG #12 wire or larger, but for my purposes, AWG #14 wiring can handle anything the solar panels throw at it. Don't forget to beef up the wiring on the protoboard to the current sensor if you do go higher than 1 A.

Figure 13.5 shows the Fritzing diagram for the project. Since there are several data points that we'll be monitoring, we'll use a Nokia 5110 LCD display to show our charging information. With the backlight LEDs off, the Nokia 5110 also has a very low current drain of about 10 mA, making it the

Anderson Powerpole Connectors

Anderson Powerpole connectors are stackable genderless single conductor housings used for power interconnections. Available in four basic housing sizes, Powerpoles can be used for quick-disconnect power connections up to 350 A. The most common Powerpole housing used in Amateur Radio applications is the PP15-45 shown in the accompanying photo. This connector housing can use 15, 30 or 45 A contacts in the same standard size housing. The Powerpole housings are designed to stack on any side of the housing, allowing you to link multiple Powerpole housings together in a wide array of configurations to suit your interconnection needs. Powerpole connectors allow you to standardize all your power connections into a single standard stacked connector configuration, allowing for quick installation and removal of your equipment and power sources.

Powerpole housings are constructed in such a way that when stacked, it is virtually impossible to connect them backward. Powerpole connectors come as a two part assembly, the housings and the contacts. The housings are available in a wide variety of colors and all accept a standard size contact. The wire is crimped or soldered to the contact and the contact is then inserted into the housing. Multiple housings can be stacked together to create a standard cable interconnect. The contacts can be removed from the housing and inserted into a different housing if needed.

I use Powerpole connectors for all my high current connection applications. All of my rotator controllers have a short pigtail connector to adapt the screw lug connections on the back of the rotor controller to a standard Powerpole connector stack, allowing quick and easy connection and disconnection of the rotator controller when it's needed for Field Day and other portable events. Because the same Powerpole connector stack orientation is used, all my CDE/Hy-Gain rotator controllers are interchangeable with my other CDE/Hy-Gain controllers as well as the CDE/Hy-Gain rotator controllers of other members of our club. This comes in real handy at Field Day when quick and easy setup is the name of the game.

Some of the more modern power supplies, such as MFJ's 4125P and the West Mountain Radio 4005i series, provide Powerpole connector outputs. Powerpole connectors are also available in chassis mount versions and in power distribution/power splitter configurations, allowing all your Powerpole outfitted equipment to use a single standard connector to connect to the power source. MFJ and West Mountain Radio, as well as other suppliers offer fuse-protected Powerpole distribution panels to handle just about every dc power distribution need you may have.

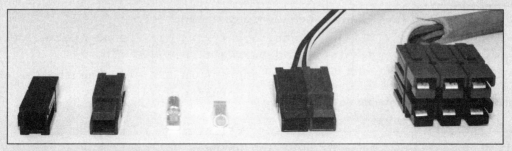

Anderson Powerpole connectors can be used in many configurations.

ideal choice for solar-powered projects. The two 7.5 V solar panels are wired in series to provide up to 15 V. In practical use, the output voltage rarely exceeds 14 V in average sunlight.

A 1N5819 Schottky diode is used as a blocking diode to prevent current from the battery from flowing back into the solar array. The 1N5819 diode was chosen due to its low forward voltage drop (0.2 V) and 3 A current handling capability. Since you'll want just about everything you can get out of the solar array, you would really prefer to have the 0.2 V forward voltage drop of a Schottky diode over the 0.7 V drop of a standard silicon diode.

You will notice that there are two sets of voltage divider resistor networks in the circuit. These are to reduce the solar array and battery voltages to levels that won't damage the Arduino analog inputs. These divider networks will reduce the voltage to approximately 25% of its actual value. Once the circuit is operational, we can calibrate the analog-to-digital conversion values with a digital voltmeter.

The Sketch

Once you have the circuit breadboarded, it's time to design the sketch. Using the sketch flowchart in **Figure 13.6**, we'll want the Arduino to read the solar array voltage, the battery voltage, and the charging current. These values will be displayed and updated every 5 seconds on the Nokia 5110 display, along with an indication of whether or not charging is occurring. Since the analog voltages being read are fed to the Arduino from a resistor divider network, we will also need to define the calibration values needed to determine the actual dc voltages coming from the solar array and the battery as well as the analog voltage from the current sensor that represents the charging current. The complete sketch for this project is found in **Appendix A** and online at **www.w5obm.us/Arduino**.

First, you'll need to define the pins and calibration values we will use for the analog voltages we'll be providing to the Arduino. The calibration values are determined by actually measuring the voltages and current with a voltmeter and using these values to derive the calibration value needed to convert the raw values to the proper voltage or current. If you don't have a voltmeter handy, the values used in the sketch should do just fine.

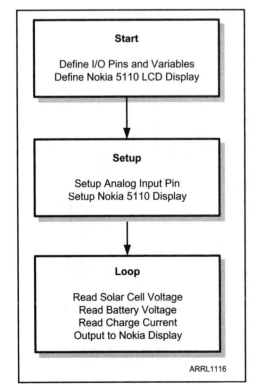

Figure 13.6 — Solar Charge Monitor flowchart.

```
#define Amps A0   // Define the Analog Input pin for the current sensor
#define Solar_In A1  // Define Analog Input pin for the Solar Cell Voltage
#define Battery_In A2  // Define the Analog Input pin for the Battery Voltage

// Define calibration value used to map the solar cell and Battery voltages
#define calibration_value 2410
```

```
// Define the calibration value used to map the charging current to milliamps
#define amp_calibration 5120
```

Next, we'll define the Nokia 5110 LCD display and define the `Small Font` for the display:

```
#include <LCD5110_Basic.h>   // Use the Nokia 5110 LCD Library
/*
 It is assumed that the LCD module is connected to
 the following pins.
      CLK - Pin 12
      DIN - Pin 11
      DC  - Pin 10
      CE  - Pin 8
      RST - Pin 9
*/
LCD5110 glcd(12,11,10,8,9); //  Assign the Nokia 5110 LCD Pins

extern uint8_t SmallFont[];   // define the Nokia Font
```

The last part of the initialization is used to define the variables holding the sensor data values:

```
// Variables to contain the solar cell and battery converted values
float solar_voltage, battery_voltage;

// Variables to contain the sensor data
int charge, solar, battery, charge_current;
```

In the `setup()` loop, the Nokia display is initialized and a brief startup message is displayed to let you know the Solar Charge Monitor is ready for action. After the startup message, the display is set up to display the template for the data values:

```
glcd.InitLCD(70);   // Initialize Nokia 5110 Display, set Contrast to 65
glcd.setFont(SmallFont);   // Set the Font to Small Font
glcd.print("KW5GP", CENTER, 0);   // Display the Startup screen
glcd.print("Solar Cell", CENTER, 8);
glcd.print("Charging", CENTER,16);
glcd.print("Monitor", CENTER,24);
delay(3000);
glcd.clrScr();   // Clear the LCD screen
glcd.print("Status", CENTER, 0);   // Set up the LCD screen for the data
glcd.print("Solar: ",0,16);
glcd.print("Battery: ",0,24);
glcd.print("Current: ",0,32);
```

In the main sketch `loop()`, we'll read the analog voltages and use a `map()` statement to convert them into the values we'll send to the LCD. The calibration values we defined at the beginning of the sketch are used here to set the voltage range in hundredths of a volt for the solar array and battery voltages, and in milliamps for the charging current.

```
// Read the sensors and display the voltages and current
charge = analogRead(Amps);   // Read the current sensor
solar = analogRead(Solar_In);   // Read the Solar cell voltage divider
battery = analogRead(Battery_In);   // Read the Battery voltage divider

// Map the Solar Cell A/D value to voltage
solar_voltage = map(solar,0,1023,0,calibration_value);

// Map the Battery A/D value to voltage
battery_voltage = map(battery,0,1023,0,calibration_value);

// Map the Current Sensor A/D value to milliamps
charge_current = map(charge,0,1023,0,amp_calibration);
```

Next the voltage and current values are displayed on the Nokia LCD. The `LCD5110_Basic.h` library has functions to print floating values to the desired number of decimal points and integers of a fixed length, allowing you to format the data nicely on the display without having to manually calculate and add spaces to have all the values line up.

```
// Display the solar voltage in 1/100 of a Volt
glcd.printNumF(solar_voltage/100,2,48,16);

// Display the battery voltage in 1/100 of a Volt
glcd.printNumF(battery_voltage/100,2,48,24);

// Display the charging current in milliamps
glcd.printNumI(charge_current,48,32,4);
```

Finally, a message is displayed on the LCD to indicate whether or not the solar array is charging the battery:

```
if (charge_current > 0)   // Charging Status Indicator
{
  glcd.print("  Charging  ",CENTER, 40);
} else {
  glcd.print("Not Charging", CENTER,40);
}
```

Figure 13.7 shows a sample display during operation. Once the circuit was fully functional and the sketch debugged, the schematic was created from the breadboard wiring (**Figure 13.8**), the project was soldered up on an Arduino

Figure 13.7 — Close-up view of the Solar Charge Monitor display.

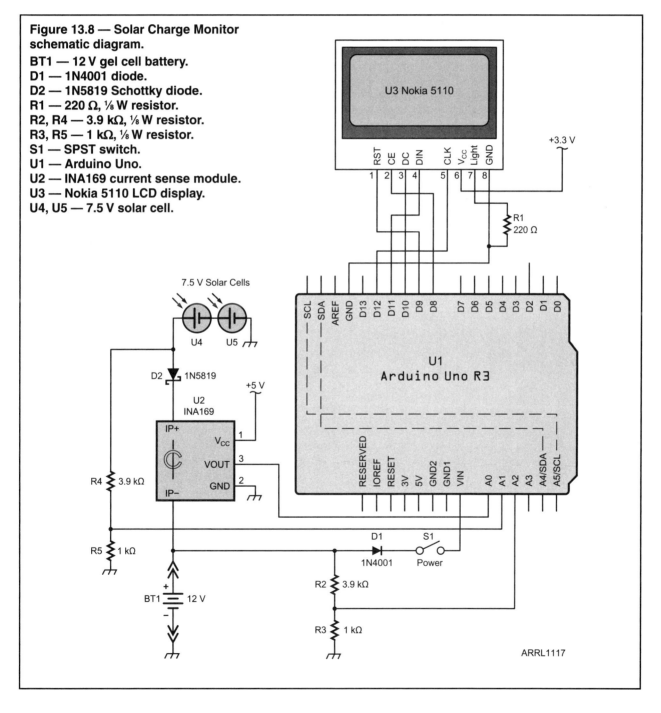

Figure 13.8 — Solar Charge Monitor schematic diagram.
BT1 — 12 V gel cell battery.
D1 — 1N4001 diode.
D2 — 1N5819 Schottky diode.
R1 — 220 Ω, ⅛ W resistor.
R2, R4 — 3.9 kΩ, ⅛ W resistor.
R3, R5 — 1 kΩ, ⅛ W resistor.
S1 — SPST switch.
U1 — Arduino Uno.
U2 — INA169 current sense module.
U3 — Nokia 5110 LCD display.
U4, U5 — 7.5 V solar cell.

Solar Battery Charge Monitor

protoshield and moved into a Solarbotics Mega Arduino S.A.F.E. enclosure. Under actual indoor conditions, the charging current after powering the Arduino runs about 80 mA — enough to slowly charge the battery.

Enhancement Ideas

You've already seen a glimpse of one of the possible enhancement ideas for this project. It is my ultimate plan to add an additional two sets of solar panels and add servo tracking capability using the pan/tilt servo head with phototransistors mounted on the four corners of the solar array to locate and track the sun across the sky to maintain the maximum charging current. The Arduino would measure the differences between the four phototransistors and automatically move the servos for maximum output.

You could also replace the INA169 current sensor with the newer INA219 version. The INA219 communicates with the Arduino using the I²C interface and has program-selectable internal gain settings to select the desired current sensing range. Since the INA219 uses the I²C bus rather than an Arduino analog input pin, you can have up to four INA219 current sensors in your solar-powered projects, allowing you to monitor all aspects of your circuits. The INA219 uses a 12-bit analog-to-digital converter to translate the current into a digital value send to the Arduino over the I²C bus. The INA219 also has an internal voltage sensor, saving you additional analog input pins.

You could also add a day/night sensing capability using phototransistors to turn off the Solar Charging Monitor when the sun goes down to reduce battery drain and turn it back on when the sun comes back up. There's also no reason that this project could not be scaled up to control a much larger, higher capacity solar array. All you would need to add is a charging current regulator circuit and geared stepper motor or worm gear-driven motor to do the positioning on a larger solar array.

Finally, to make the project even more portable and energy efficient, you can replace the Arduino Uno with a smaller Arduino, such as a Nano, and mount everything in the base of the solar panel assembly. Using a Nano instead of the Uno would save you about 30 mA of charging current, giving you just that much more charging current to the battery.

I'm really looking forward to using this project to charge and power the alternative power QRP station we use at ARRL Field Day and I think it would be neat to watch the solar panels track the sun across the sky all on their own.

References

Adafruit Industries — **www.adafruit.com**
Anderson Power Products — **www.andersonpower.com**
Henning Karlsen — **www.henningkarlsen.com/electronics/**
Texas Instruments — **www.ti.com**
Solarbotics — **www.solarbotics.com**
SparkFun Electronics — **www.sparkfun.com**

CHAPTER 14

On-Air Indicator

The finished On-Air Indicator mounted in a SparkFun project enclosure.

One of the fun things you'll discover about the Arduino is its versatility. It's like having an adult Erector Set. You can build something, make it work, take it apart, rearrange the pieces, and you've got something entirely different. Such is the case with the On-Air Indicator. Using some of the same pieces we built for the RF Probe in Chapter 12, we can create an entirely different project with an entirely different purpose.

As I mentioned back when we built the CW Beacon and Foxhunt Keyer, my friend and fellow Arduino builder, Tim Billingsley, KD5CKP, operates a finicky 10 meter beacon. Sometimes it just goes dead without warning. Until we can dig into things and see what exactly is going on, Tim wanted a better way to tell if the beacon was actually transmitting without having to leave a power/

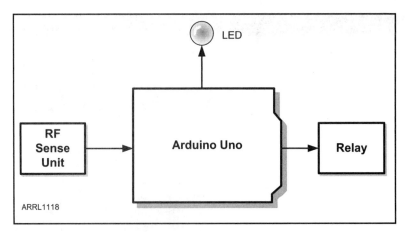

Figure 14.1 — On-Air Indicator block diagram.

SWR meter attached to it, looking at the beacon transmitter's meter, or leaving a receiver tuned to the beacon frequency all the time. With just a few twists and a slightly different circuit design, we can turn the RF Probe from Chapter 12 into an On-Air Indicator. Like Tim, you can use it to verify that your beacon is transmitting, or you can use the relay output to light a sign on your shack door telling others that you're busy chasing some DX and to enter quietly, or slide your dinner under the door.

Unlike the RF Probe project, we're not really interested in the signal strength — we're more concerned with whether or not there is a signal. Since we're only looking for an on or off indication, as the block diagram (**Figure 14.1**) shows, we've replaced the bar graph LED with a single LED and a relay output to indicate when an RF signal is present.

At the heart of this project is the exact same RF sensing unit we built for the RF Probe project in Chapter 12. This time around, we'll use the RF sensing unit to determine the presence of an RF signal rather than read the signal's strength, and we'll have the Arduino read the probe and trigger the LED and relay if the signal strength exceeds a threshold value.

If you haven't already built the RF sensing assembly, you can find the parts list and instructions for building one in Chapter 12. Now you can see the benefits of using standard connectors for all the pieces of your projects. If you built the RF Probe project, all you have to do is disconnect the RF sensing unit from the RF Probe project, connect it to the Arduino protoshield used on this project and you're almost ready to roll.

Figure 14.2 shows the Fritzing diagram for the On-Air Indicator. You can see where we've replaced the bar graph LED from the RF Probe project with an LED and a relay driven by a 2N2222A transistor. Whenever you use a relay in this manner, be sure that the relay has an internal clamping diode, or add one externally to prevent reverse voltage from damaging the transistor when the relay is de-energized. As a general rule, I always include a 1N4001 clamping diode (D2) in the circuit design and leave it out if the relay I end up using has an internal clamping diode.

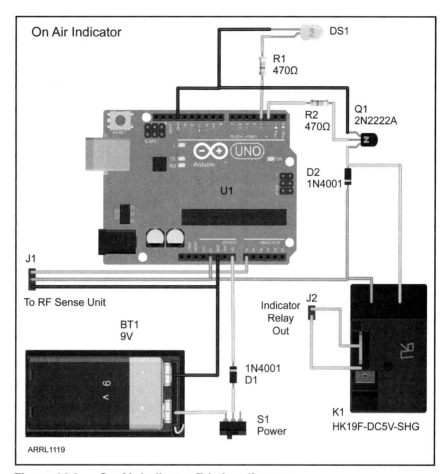

Figure 14.2 — On-Air Indicator Fritzing diagram.

The Sketch

Figure 14.3 shows the flowchart for the sketch used in this project. With projects as straightforward as the On-Air Indicator, you may not need a flowchart, but I have found that it helps to flowchart every sketch. It's a good habit to get into and helps you stay on track as you write the sketch. With a flowchart, even if your sketch building is interrupted, you can use the flowchart like a checklist and pick right back up where you left off. In keeping with the Erector Set theme, the sketch for the On-Air Indicator is based on the sketch we used in the RF Probe project.

Now that you have the circuit wired up and ready to go, it's time to write the sketch itself. Starting out, we'll define the I/O pins and signal threshold, along with the LED delay and relay hold delay. You may have noticed that the Arduino I/O pins are usually defined at the start of the sketch. This is done to give the I/O pin a name that we can remember. It's a whole lot easier to remember that digitalWrite(relay,

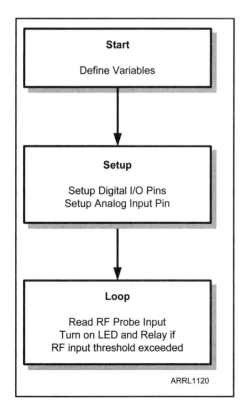

Figure 14.3 — On-Air Indicator flowchart.

HIGH) turns on the relay as compared to the `digitalWrite(2, HIGH)` statement we would have to use if we had not defined the I/O pins. For more complex sketches, having all the I/O pins defined makes keeping track of what's going on in the sketch a whole lot easier. Our design calls for the On-Air Indicator to turn on the LED and relay when the RF signal detected by the RF sensing unit exceeds a desired threshold. The LED and relay will remain energized for a specified delay time when no signal is detected.

```
#define Signal_Threshold 100  // The signal strength required to turn on
#define Signal_Pin A0    // define the Analog Input pin for the RF sensor
#define relay 2  // define the Relay control pin
#define LED 3  // define the LED pin
#define hold_time 2000  // Hold for 2 seconds after signal input goes away
```

We'll also need to define a few variables to keep track of things in the sketch. Here we've introduced the Boolean variable type. Boolean variables are used to indicate an on or off, high or low, true or false binary state. Boolean variables use less memory than other variables and work well in simple yes/no decision points within your sketch. You will also note that the variable type for the timeout value is an unsigned long integer. This is needed since we will be using the Arduino `millis()` function to determine when to turn off the LED and relay. The `millis()` function returns an unsigned long integer value, so we will need to use variables with the matching variable type to be able to calculate the turn-off time.

```
int Signal_Strength = 0;  // stores the value coming from the RF sense unit
unsigned long timeout = 0;  // stores the timeout time
bool on_air = false; // indicates that the on air indicator is active
```

The `setup()` loop for the On-Air Indicator is used to set the pin mode for the LED and relay control pins and to make sure they are turned off before the main `loop()` begins. As you can see, using the defined names for the pins rather than the pin number makes it much easier to follow what the sketch is doing.

```
pinMode(relay, OUTPUT);   // set the relay control pin
pinMode(LED, OUTPUT);   // set the LED pin
digitalWrite(relay, LOW);   // Turn off the relay and LCD
digitalWrite(LED, LOW);
```

In the main `loop()`, we read the analog signal strength and if the value exceeds the predefined threshold, the LED and relay are turned on. The hold delay is then added to the current value of `millis()`. As long as the signal strength exceeds the threshold value, the delay timer continues to be added to the current `millis()` value.

When the signal strength drops below the threshold, the sketch continues to check for the current `millis()` value to exceed the time saved in the timeout variable. When the current time exceeds the stored value, the LED and relay are turned off.

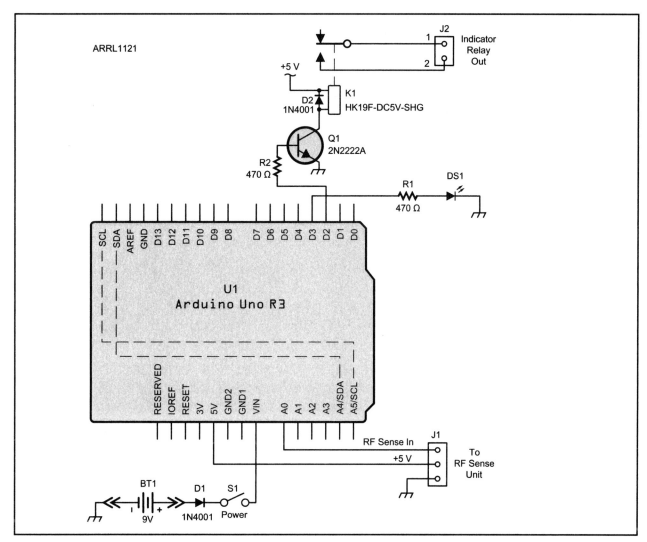

Figure 14.4 — On-Air Indicator schematic diagram.

BT1 — 9 V battery.
D1, D2 — 1N4001 diode.
DS1 — 5 mm LED.
J1 — 3-pin header with cable and plug to match RF sensing unit.
J2 — Jumper header or builder's choice of jack.

K1 — 5 V relay (HK19F-DC5V-SHG).
Q1 — 2N2222A transistor.
R1, R2 — 470 Ω, 1/8 W resistor.
S1 — SPST switch.
U1 — Arduino Uno.

```
// Read the Signal Strength from the RF Sensor
Signal_Strength = analogRead(Signal_Pin);

// Turn on relay and LED if threshold is exceeded
if (Signal_Strength >= Signal_Threshold)
{
  digitalWrite(relay, HIGH);
  digitalWrite(LED, HIGH);
  timeout = millis() + hold_time;  // set the timeout time
  on_air = true;
}

// If we've passed the timeout time, turn off the relay and LED
if (millis() > timeout && on_air)
{
  digitalWrite(relay, LOW);
  digitalWrite(LED, LOW);
  on_air = false;
}
```

Once the project and sketch were debugged and working on the breadboard, the schematic (**Figure 14.4**) was used to build the Arduino protoshield and the finished project was mounted in a SparkFun Arduino project case. **Figure 14.5** shows the finished project.

Enhancement Ideas

You can turn this project into a wireless On-Air Indicator if you were to use an Arduino Nano and mount all the components, including the RF sensing

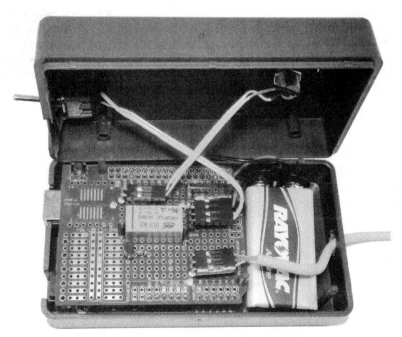

Figure 14.5 — Inside view of the On-Air Indicator.

unit inside an LED lightbox with a clear front plate and a mask for the On-Air lettering. You could power everything with batteries, use a small antenna for RF sensing extending outward from inside the box, and hang it on the wall outside your shack whenever you're operating.

References

SparkFun Electronics — **www.sparkfun.com**

K. Austermiller, KB9STR, "An RF Driven On-Air Indicator," *QST*, Aug 2004, pp 56-57.

J. Noakes, VE7NI, "The 'No Fibbin' RF Field Strength Meter," *QST*, Aug 2002, pp 28-29; Feedback Sep 2002 *QST*, p 88.

CHAPTER 15

Talking SWR Meter

The finished Talking SWR Meter mounted in a Solarbotics Arduino Mega S.A.F.E.

One of the fun things about the Arduino is that there are more and more new devices and modules available for it all the time. This project was originally designed to be just a standard digital SWR meter. While ordering some of the enclosures I was planning to use for my projects, I came across the Emic 2 text-to-speech module. Suddenly a whole range of new project ideas came to mind.

The first thought was my typical, "How cool would it be to have an SWR meter that could speak and tell you what the SWR is?" This was quickly followed by the realization that the Emic 2 text-to-speech module would open up a whole world of new projects for operating in low light conditions, and for visually impaired hams. Several of my ham friends are visually impaired and now I had ideas for projects they could build (or have someone build for them) and get some serious usefulness out of them. Also, since many of our club's operating events are at night with the usual dim lighting, the idea of an SWR meter that could speak the SWR had my complete attention.

The Emic 2 Text-to-Speech Synthesizer

The Parallax/Grand Idea Studios Emic 2 shown in **Figure 15.1** is not your run-of-the-mill text-to-speech synthesizer. In addition to converting standard ASCII text to speech, the Emic 2 natively converts the text into English or Spanish and has nine pre-defined voice styles, along with dynamic control of various speech characteristics such as pitch, speaking rate, and word emphasis. All of these features can be controlled from within your sketch and changed

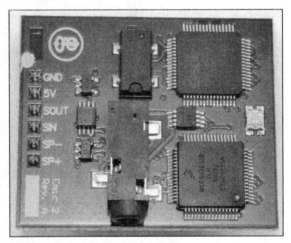

Figure 15.1 — The Parallax/Gran Idea Studios Emic 2 text-to-speech module.

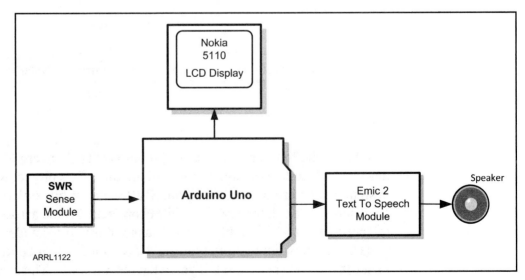

Figure 15.2 — Talking SWR Meter block diagram.

on the fly. The Emic 2 communicates with the Arduino using a standard TTL serial port at 9600 baud and works well with the Arduino `Software Serial` library, allowing you to use any two digital I/O pins to connect the Emic 2 to the Arduino. The Emic 2 is a fully self-contained module that includes a 300 mW on-board audio power amplifier and a standard 3.5 mm audio jack. The output volume can also be controlled from within your sketches. **Figure 15.2** is a block diagram of the project.

SWR Sense Head

Now that we know what we're going to do with the SWR data when we get it, we have to figure out a way to read the SWR and pass that information on to the Arduino. Remember in the Introduction when I told you how handy *The ARRL Handbook* can be? As it turns out, others have passed this way ahead of us and left some very useful breadcrumbs. Recent editions of *The ARRL Handbook* have a section on building a microprocessor controlled SWR monitor.[1] While the circuit design in the *Handbook* is for a Parallax "Propeller" microprocessor, the SWR sensing head used in the *Handbook* project is perfect for our needs. **Figure 15.3** shows the circuit.

The SWR sense head described in the *ARRL Handbook* is based on "The Tandem Match — An Accurate Directional Wattmeter" by J. Grebenkemper,

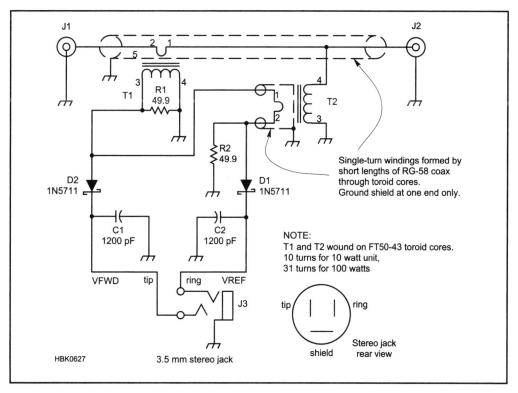

Figure 15.3 — SWR sense head schematic from *The ARRL Handbook*.
C1, C2 — 1200 pF, 50 V capacitor.
D1, D2 — 1N5711 Schottky diode.
J1, J2 — SO-239 coax jack or builder's choice.
J3 — ⅛ inch mini stereo phone jack.
L1, L2 — FT50-43 toroid core; 10 turns for 10 W unit, 31 turns for 100 W unit.
R1, R2 – 49.9 Ω, ¼ W resistor.
Radio Shack (P/N 270-238) aluminum project enclosure

KA3BLO.[2] Using a simple and easy-to-construct design, the SWR sense head will output an analog voltage representation of both the forward and reflected power simultaneously. We can supply these voltages to the Arduino's analog inputs and directly read them.

I built my SWR sense head using the exact circuit design in the *2013 ARRL Handbook* on a small piece of perfboard mounted inside a Radio Shack (P/N 270-238) aluminum project enclosure to provide RF shielding. The *Handbook* project used surface-mount components, but I used leaded parts in my version. In the *Handbook* design, you have the option of building the unit for sensing 10 W or 100 W simply by changing the number of turns of wire on the two toroids used to transfer a small amount of RF energy into the SWR sensing circuitry. For this project, I chose to build the 10 W version, but there would not be any issues at all if you chose to build the 100 W version for your project. **Figures 15.4** and **15.5** show my completed SWR sense head.

Figure 15.4 — Inside view of the SWR sensing unit.

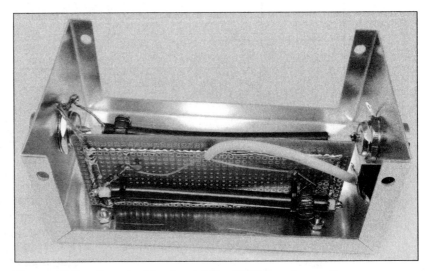

Figure 15.5 — Rear inside view of the SWR sensing unit.

Arduino and Related Hardware

Now that we have all the parts and pieces ready, we can use the Fritzing diagram (**Figure 15.6**) to start breadboarding up the Arduino side of things. Whatever you do, don't forget to include the 5.1 V Zener diodes (D2 and D3) on the SWR sense head inputs to the Arduino. These diodes are there to prevent any SWR signal input voltages from exceeding 5 V and potentially damaging the analog inputs of the Arduino. Also, notice that the Emic 2 module is not connected to the standard Arduino serial pins 0 and 1. We will be using the Arduino `Software Serial` library to communicate with the Emic 2 on digital I/O pins 2 and 3.

Why Use the Software Serial Library?

You may be asking why this project is using the `Software Serial` library to communicate with the Emic 2 when the Arduino already has a

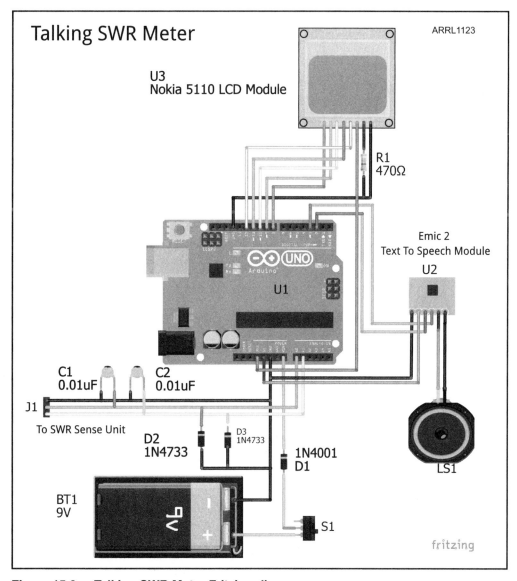

Figure 15.6 — Talking SWR Meter Fritzing diagram.

perfectly usable serial port on digital I/O pins 0 and 1. The reason we use the `Software Serial` library is that the Arduino shares the USB port used to connect to your workstation with the hardware serial port on digital I/O pins 0 and 1. This means that when you have the Arduino connected to your workstation and you're uploading sketches or using the `Serial Monitor` in the IDE, you can't use the hardware serial port to talk to other devices. The easy way around this is to use the `Software Serial` library and attach the Emic 2 or other serial device to any two digital I/O pins of your choosing. The Arduino Leonardo has addressed this issue, and actually has two hardware serial ports onboard. You could use an Arduino Leonardo in this project instead of an Uno if you would prefer to use a hardware serial port instead of the `Software Serial` library. Since the Uno and its variants are currently the most common Arduinos, we'll use the Uno and the `Software Serial` library for this project.

The Sketch

With the circuit ready on the breadboard, we can start putting together the sketch. While this may seem like a somewhat complicated project, using the flowchart in **Figure 15.7** we can break it down into bite-size pieces. For me, that is the key to writing Arduino sketches. When you look at them as a whole, they may seem difficult and complex. When you break the sketch down into

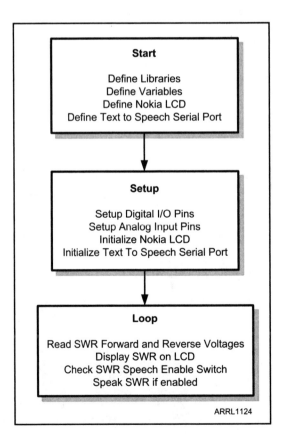

Figure 15.7 — Talking SWR Meter flowchart.

smaller pieces, you soon realize that it really isn't that difficult at all. They key is to plan things out ahead of time, and you can use your flowchart like a map to stay on course. You can find the complete sketch and libraries for this project in **Appendix A** and online at **www.w5obm.us/Arduino**.

In the initialization part of the sketch, we start by defining the Nokia 5110 LCD display and the `Software Serial` libraries, assigning the I/O pins, and initializing the library objects:

```
#include <LCD5110_Basic.h>  // Include Nokia 5110 LCD Library
LCD5110 glcd(12,11,10,8,9);  // Define the LCD object
extern uint8_t SmallFont[];  // Define the LCD Font

// include the SoftwareSerial library for the Emic 2 module
#include <SoftwareSerial.h>

#define rxPin 2    // Serial input (connects to Emic 2 SOUT)
#define txPin 3    // Serial output (connects to Emic 2 SIN)
#define audio_on 7 // Audio Enable Switch Pin

// set up a new serial port
SoftwareSerial emicSerial =  SoftwareSerial(rxPin, txPin);
```

Next, we'll define the constants and variables used in the sketch. Since we want to calculate voltage from the digital values provided by the Arduino's analog-to-digital (A/D) converter, we define the conversion value for one A/D count. Since the SWR voltages will be small and we'd like them to be as accurate as possible, we'll use the float variable type for these variables.

```
// Volts per ADC count (0 to 5 volts, 0 to 1023 A/D counts)
#define adc_count 0.0048828125

float V_Fwd, V_Ref = 0, V_SWR ;   // Define the variables
```

In the `setup()` loop, we do the usual `pinMode()` thing to set up the I/O pins for the Emic 2 and the `Audio Enable` switch on pin 7. Notice that we are also performing a `digitalWrite()` to the `Audio Enable` switch pin while it is configured as an input. This enables the internal 20 kΩ pull-up resistor on that I/O pin, saving us from having to use an external pull-up resistor on the `Audio Enable` switch pin.

```
pinMode(rxPin, INPUT);
pinMode(txPin, OUTPUT);
pinMode(audio_on, INPUT);
digitalWrite(audio_on, HIGH); // Enable internal 20K pullup resistor
```

Next, we'll start the `Software Serial` port and configure it for 9600 baud, which is the default baud rate used by the Emic 2. We'll also start the Nokia LCD and show a brief startup message so we can tell that things are running. You can also set the Nokia 5110 display contrast by including the contrast value as a parameter in the `glcd.InitLCD()` command. The default setting for the contrast is 70, however I have found that there is a wide variation in the contrast between display modules, and that contrast also varies significantly with the voltage supplied to V_{cc} on the Nokia 5110 LCD.

```
emicSerial.begin(9600);    // set the data rate for the SoftwareSerial port

glcd.InitLCD(60);    // Initialize the Nokia 5110 Display, set contrast to 60
glcd.setFont(SmallFont);

// Display the Startup screen
glcd.clrScr();
glcd.print("KW5GP", CENTER,0);
glcd.print("SWR Meter",CENTER,8);
delay(3000);
glcd.clrScr();
```

In the final portion of the `setup()` loop, we'll initialize the Emic 2 module, select the voice we want to use, and set the volume level. Note that if your sketch seems to hang at this point, check the connections to the Emic 2, as the sketch will not continue until it receives the colon (:) character from the Emic 2 to indicate that it is online and ready to accept data.

```
emicSerial.print('\n');     // Send a CR in case the system is already up

/* When the Emic 2 has initialized and is ready, it will send a single ':'
character, so wait here until we receive it
/*
while (emicSerial.read() != ':');
delay(10);                          // Short delay
emicSerial.println("n1");    // Set voice to Voice 1
delay(500);
emicSerial.print('v');    // Set the volume to +18db
emicSerial.println("18");
emicSerial.flush();                 // Flush the receive buffer
```

Now we're ready to start the main `loop()`. First we read the forward and reverse voltages coming in on the Arduino analog input pins. The values are then converted to their actual voltages and displayed on the Nokia LCD.

```
glcd.print("Fwd:",0,0);  // Display the SWR information on the LCD
glcd.print("Ref:",0,8);

// Read the analog inputs and convert them to voltages
V_Fwd = analogRead(0) * adc_count;
V_Ref = analogRead(1) * adc_count;

// display the Forward and Reflected voltages
glcd.printNumF(V_Fwd,7,30,0);
glcd.printNumF(V_Ref,7,30,8);
```

Next we'll calculate and display the SWR. As an error check, if the reflected voltage is higher than the forward voltage, the SWR is reported as 0. To protect your equipment, don't transmit without either an antenna or dummy load connected to your radio. Also, you should not transmit when the SWR at the transmitter exceeds 3:1 unless your transmitter has an internal tuner and can adjust the SWR to a safe level. Many modern radios also have SWR protection circuits to limit the transmitter power output when high SWR is detected, but that is not a good reason to tempt fate. If your SWR exceeds 3:1, troubleshoot and repair the problem, and do not transmit at full power until you have the SWR within a safe range for your transmitter.

For my testing and calibration, I used a TEN-TEC Rebel Model 506 QRP CW transceiver with the output set to 3 W, and compared the SWR readings into a 50 Ω dummy load and my GAP Challenger HF vertical antenna with a Diamond SX-600 SWR/power meter. The SWR readings between the Diamond SX-600 and my homebrew SWR sense head were within 5% of each other on both 20 and 40 meters, and no adjustments or modifications were needed on the SWR sense head.

```
// Calculate VSWR
if (V_Fwd > V_Ref)
{
  V_SWR = (V_Fwd + V_Ref) / (V_Fwd - V_Ref);
} else {
  V_SWR = 0;  // Display SWR of 0 to 1 if Reflected greater than Forward
}

// Display the VSWR
glcd.print("SWR: ",0,24);
glcd.print("           ",30,24);
glcd.printNumF(V_SWR,1,30,24);
glcd.print(" : 1",56,24);
```

Now for the fun stuff. We'll check to see if the Audio Enable switch on digital I/O pin 7 is turned on. If it is, we'll have the Emic 2 speak the SWR for us.

```
if (digitalRead(audio_on) == LOW)
{
  // Audio is enabled - speak the SWR

  // Send the Emic 2 the command to speak the text that follows
  emicSerial.print('S');

  // Send the desired string to convert to speech
  emicSerial.print("S W R is ");
  emicSerial.print(V_SWR,1);
  emicSerial.print(" to 1");

  // Send the Emic 2 the New Line character to signify end of text
  emicSerial.print('\n');

  /*
  Wait here until the Emic 2 responds with a ":" indicating it's ready to
  accept the next command
  */
  while (emicSerial.read() != ':');
}
```

Once the circuit and sketch were tested and debugged, the finished project was moved onto an Arduino protoshield using the schematic diagram in **Figure 15.8**. The completed project was mounted inside a clear Solarbotics Mega S.A.F.E. enclosure to protect the Arduino and internal components from the elements.

Enhancement Ideas

With proper shielding, you could use a smaller Arduino such as a Nano, and house the entire project inside the SWR sense head project box. The Emic 2 text-to-speech module opens up a wide variety of enhancement options, such as using the module to provide an alert when the SWR exceeds a preset level. We had hours of fun coming up with phrases we would have the SWR meter speak when it encountered alarm conditions, most of which cannot be repeated here. You could also add an RGB LED to indicate the SWR status for those times when you don't want the distraction of the speech module. Finally, it would be ideal to add the forward and reflected transmitter output power to the display and have the Arduino speak the power as well. All you would need to do is calibrate the forward and reflected SWR voltages to a known-good power meter, calculate the power based on the readings, and display them on the Nokia LCD.

References

Diamond Antennas — **www.diamondantenna.net**
GAP Antenna Products — **www.gapantenna.com**
Grand Idea Studio — **www.grandideastudio.com**
Henning Karlsen — **www.henningkarlsen.com/electronics/**
Parallax — **www.parallax.com**

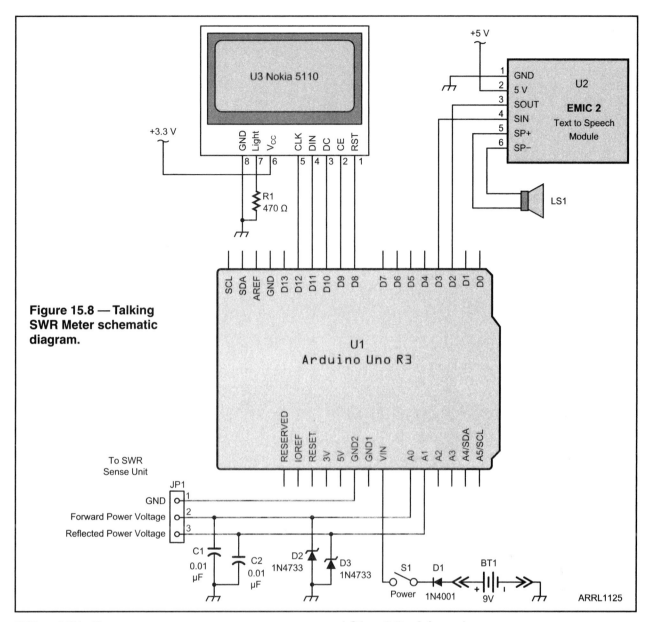

Figure 15.8 — Talking SWR Meter schematic diagram.

BT1 — 9 V battery.
C1, C2 — 0.01 µF, 35 V ceramic capacitor.
D1 — 1N4001 diode or equivalent.
D2, D3 — 1N4733 5.1V Zener diode.
J3 — Jumper header or builder's choice of jack to match SWR sense head.

LS1 — 8 Ω mini speaker.
S1 — SPST toggle switch.
U1 — Arduino Uno.
U2 — Emic 2 text-to-speech module.
U3 — Nokia 5110 LCD display module.

RadioShack — **www.radioshack.com**
Solarbotics — **www.solarbotics.com**
TEN-TEC — **www.tentec.com**

Notes

[1]"A Microprocessor Controlled SWR Monitor" by Larry Coyle, K1QW, appears in Chapter 24, Station Accessories, of the 2010 to 2014 editions of *The ARRL Handbook for Radio Communications*. In my 2013 edition the project is on pages 24.4 – 24.9.

[2]J. Grebenkemper, KA3BLO, "The Tandem Match — An Accurate Directional Wattmeter," *QST*, Jan 1987, pp 18-26.

CHAPTER 16

Talking GPS/UTC Time/ Grid Square Indicator

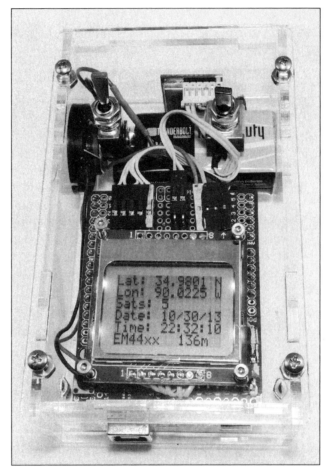

The finished GPS/UTC Time/Grid Square Display mounted in a Solarbotics Arduino Mega S.A.F.E.

A Global Positioning System (GPS) receiver/display can do more than just tell you latitude, longitude, and how to get where you're going. Since a GPS uses highly accurate timing signals to determine your location, it can also provide very accurate time information. In fact, there is a whole series of National Maritime Electronics Association (NMEA) messages, or "sentences" that use the timing information provided by the GPS satellites, and that information is processed by a GPS receiver into usable information. Using these NMEA sentences, we can extract all kinds of useful information from our GPS receiver.

So what can we do with all this GPS information? Up until recently, I was asking myself the same thing. I've had a GPS unit in my car for years, and aside from that one time it sent me down a dirt road out into the middle of a corn field, it's been very handy and reliable. At a recent Arduino presentation, I was asked to create several new projects for the Arduino and introduce them during the presentation. One of these projects was a simple UTC/local time clock using a real time clock-calendar module. This worked well, but you had to set the time manually and while the clock was reasonably accurate, it would still drift and lose a few seconds of accuracy over time.

Maidenhead Grid Locators

What are these "grid squares" you keep hearing about? The Maidenhead Locator System was devised in 1980, at a meeting of European VHF managers to create a global geographic coordinate system that Amateur Radio operators could use to denote their location. Named after the town in England where the meeting was held, Maidenhead grid locators divide the world into small areas, measuring 1° of latitude by 2° of longitude (approximately 70 by 100 miles in the continental US). A grid square is composed of 4 to 6 characters and is measured northward starting from the South Pole and eastward starting from the anti-meridian of the Prime Meridian in Greenwich, England (somewhere in the middle of the Pacific Ocean).

The first two letters of a grid square, also known as the field, are used to denote 18 separate 20° blocks of longitude and latitude, donated by the letters A thru R. The first letter is the longitude and the second letter is the latitude. The field is followed by a pair of numbers, known as the square. These numbers range from 0 thru 9 each and divide the field into "squares" of 1° latitude by 2° longitude, with the first number being longitude and the second number being latitude. Optionally, the grid square can be further divided into subsquares of 2.5 minutes of latitude by 5 minutes of longitude (approximately 3 by 4 miles in the continental US), denoted by a pair of letters a-x, with the first letter representing longitude and the second letter representing latitude.

As an example, my longitude is 90.0223° W. This places me within the fifth block of 20° of longitude from the anti-meridian, so the first letter of my grid square is "E". My latitude is 34.9799° N, which places me within the thirteenth block of 20° from the South Pole, so the second letter of my grid square is "M". Within the field, the calculations show that I am in the fifth 2° block of longitude. Since the numbering system starts at zero, the first number in my square is "4". Applying the same calculation to my latitude, the calculation shows that I am also in the fifth 1° block of latitude. Therefore, my grid square is EM44. You can continue the calculations on down to the subsquare and determine that my exact grid square is EM44xx. Fortunately, the formula for calculating grid squares has already been developed for us, so there is rarely a need to calculate a grid square manually.

Precise Time Signals

Shortly thereafter, I got involved in a team effort to get the JT65-HF digital communications mode running on the new TEN-TEC Rebel Model 506 QRP CW transceiver. The Rebel is an Open Source QRP CW transceiver based on the Digilent chipKit Uno32, an 80 MHz variant of the Arduino Uno. The Rebel uses a Texas Instruments AD9834 direct digital frequency synthesis (DDS) integrated circuit to generate the transmitted signal as well as the IF mixing signal for the receive side of the Rebel. Through the use of some serious magic and some absolutely amazing coding by Joe Large, W6CQZ, the team came up with a method to actually command the chipKit Uno32 in the Rebel to shift the AD9834 DDS transmit frequency on the fly to generate the 64 tones necessary for JT65.

Since the Rebel is an ideal portable QRP transceiver that can go with you just about anywhere, we wanted a way to acquire accurate time anywhere for the JT65-HF mode to function correctly. The JT65-HF protocol requires that a transmission must start at precisely 1 second into a new minute. If you're out in the middle of nowhere with no Internet access to NTP (Network Time Protocol) time servers, your only option is to set the time manually on the workstation running the JT65-HF software. We wanted the JT65 operations on the Rebel to be totally self-sufficient and not rely on any external time source for the time synchronization. The light bulb came on, and a GPS interface was quickly built on a modified Arduino protoshield and mounted on the chipKit Uno32's shield expansion pins inside the Rebel. Our Rebel prototype can now provide accurate time and location, and as an added treat, it can also calculate and display your current Maidenhead grid locator. (Hams usually just call these "grid squares." See the sidebar, "Maidenhead Grid Locators.")

Ok, so what if you don't use JT65 or don't have a Rebel? Well, many of the contests, particularly the VHF contests, use your grid square as part of the contest exchange. If you're into mobile operating, it's difficult to keep track of your grid square as you travel down the road. It's also cool to have this project around for Field Day and other portable events, providing accurate time, altitude, and grid square information.

Skylab GPS Receiver

The Skylab SkyNav SKM53 shown in **Figures 16.1** and **16.2** is a fully self-contained GPS module that supports the standard NMEA-0183 messaging format. With its −165 dBm tracking sensitivity, the SKM53 works well even in "urban canyons" and dense foliage environments. Capable of tracking up to 22 GPS satellites, the SKM53 can provide highly accurate location, time, and altitude information. The SKM53 also provides heading and speed information, allowing it to be used for navigation in addition to providing fixed location information. The SKM53 communicates with the Arduino using a standard TTL serial port running at a default speed of 9600 baud.

Figure 16.1 — Rear view of the Skylab SKM53 GPS module.

Figure 16.2 — The Skylab SKM53 GPS module.

The NMEA-183 Protocol

The NMEA-183 protocol is an ASCII text based protocol. All messages start with a $, end with a carriage return/line feed (CR/LF), and contain a checksum to detect corrupted messages. The SKM53 GPS module supports the NMEA-183 standard GGA, GLL, GSA, GSV, RMC, VTG, and ZDA messages. Of these messages, the GGA message contains the majority of the information we'll be using in our projects. A standard GGA message will look like this:

$GPGGA, 150028.000,3458.8015,N,09001.3496,W,1,9,0.90,93.7,M,-30.0,M,,*55

This message breaks down into:

$GPGGA — GGA Message Type

150028.000 — UTC Time in hhmmss.sss
3458.8015 — Latitude in degrees and minutes ddmm.mmmm
N — North Latitude
09001.3496 — Longitude in degrees and minutes ddmm.mmmm
W — West Longitude
1 — Position Fix Indicator. A "1" indicates the reading is in GPS SPS mode and the fix is valid
9 — The number of GPS satellites used to determine the position fix
0.90 — The Horizontal Dilution of Precision (HDOP) of the fix
93.7 — Altitude above mean sea level
M — Altitude reading is in meters
-30.0 — The Geoids Separation value (correction for tidal influences)
M — Geoids reading is in meters
*55 — Message Checksum

And the answer to your next question is yes, if you plug these numbers into your GPS, you will find yourself on top of the work table in my lab, give or take a few meters. As you can see, there is a lot of valuable information packed into an NMEA sentence. Fortunately, there is a pair of excellent libraries for the Arduino that handle the decoding of the GPS messages, without us having to parse out all that data manually. The Arduino `TinyGPS` and `TinyGPS++` libraries from **www.arduiniana.org** are outstanding for any sketches needing to extract data from your GPS module. For this project, we will be using the `TinyGPS` library.

To kick things up a notch, we'll also be using the Emic 2 text-to-speech module we used in the last project to have our GPS project speak the time and our current Maidenhead grid locator. This can come in real handy for those times you're operating mobile and don't want to take your eyes off the road.

Figure 16.3 shows the block diagram for the Talking GPS/UTC Clock and Grid Square Indicator. For this project, we'll have the Arduino read the GPS,

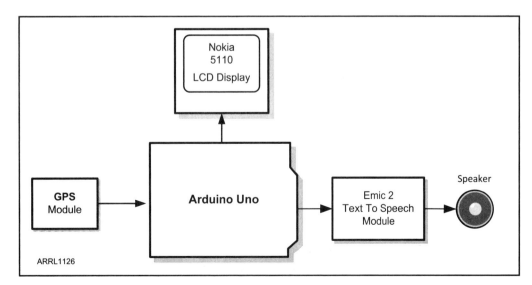

Figure 16.3 — Talking Grid Square Display block diagram.

and then extract the longitude, latitude, date, time, and altitude information. We'll have the Arduino calculate our grid square and display this on a Nokia 5110 LCD display. We'll also use the Emic 2 module to speak the time and grid square when the audio is enabled.

Now that we know what pieces we'll be using in this project, we'll create the Fritzing diagram (**Figure 16.4**), which will show us how to wire everything up. In this project, we'll be setting up two `Software Serial` ports, one for the GPS module, and the other for the Emic 2. To demonstrate how easy it is to design and create complex projects for the Arduino, this project needs only nine wires to connect to the Arduino, not including the power and speaker connections.

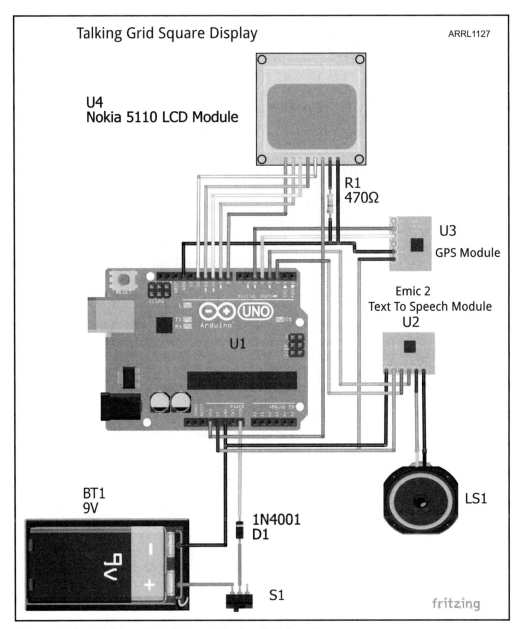

Figure 16.4 — Talking Grid Square Display Fritzing diagram.

The Sketch

With everything wired up and ready to go, we can now begin to write the sketch. Using the flowchart in **Figure 16.5**, we have all the steps laid out in the order that we need to do them. As you will see as we get into the sketch, it really is as easy as the flowchart makes it look. The libraries do most of the work for us; all we have to do is "glue" the various pieces together. The entire sketch and libraries can be found in **Appendix A** and online at **www.w5obm.us/Arduino**.

Starting off with the sketch, you will see that we include a library we haven't discussed yet. The Arduino `math.h` library is a library of extended math functions that is included with the Arduino IDE, so all we have to do is define it to gain access to its features. For this sketch, we need the ability to calculate the absolute value of a floating variable. The basic Arduino math functions do not have this ability, so we add that functionality using the `math.h` library.

```
#include <math.h>   // so we can get absolute value of floating point number
```

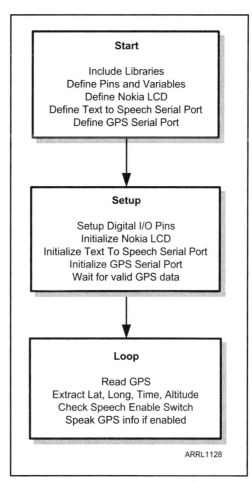

Figure 16.5 — Talking Grid Square Display flowchart.

Next, we'll include the Nokia 5110 library and define the LCD display object and font:

```
#include <LCD5110_Basic.h>   // Include Nokia 5110 LCD Library

LCD5110 glcd(12,11,10,8,9);  // Define the LCD object
extern uint8_t SmallFont[];  // Define the LCD Font
```

Now, we'll include the `Software Serial` library for the Emic 2 and GPS modules. You will note that the RX pin on the Emic 2 module and the TX pin on the GPS module are defined as -1. This tells the library that these pins will not be active when we initialize the modules later in the sketch. The reason we don't enable these pins is because the `Software Serial` library doesn't like to share time between the two modules. Since we have data coming in from the GPS constantly, we can't sit in a loop waiting for the Emic 2 to respond with its usual ":" ready for input prompt after we send it data. We'll have to assume the Emic 2 is ready and accepting data as we send it. Again, you could use an Arduino Leonardo to work around these issues and place one of the two modules on the Leonardo's hardware serial port, but since we've standardized on the Arduino Uno for now, we'll do it the hard way. In actual operation, you'll see that everything comes together and plays well with each other.

```
// include SoftwareSerial library so we can talk to the Emic 2 module
#include <SoftwareSerial.h>

// Rx set to -1 (disabled) because we don't want to receive from Emic 2
#define emic_rxPin -1    // Serial input (connects to Emic 2 SOUT)
#define emic_txPin 3     // Serial output (connects to Emic 2 SIN)
#define audio_on 7       // Audio Enable Switch on Pin 7

// set up a new serial port for the Emic 2
SoftwareSerial emicSerial =  SoftwareSerial(emic_rxPin, emic_txPin);

#define gps_rxPin 5      // Serial input (connects to GPS TxD)

// Set to -1 (disabled) because we don't need to send anything to GPS
#define gps_txPin -1     // Serial output (connects to GPS RxD)
```

Here we include the `TinyGPS.h` library, assign the GPS I/O reset pins, and define the GPS object:

```
#include <TinyGPS.h>  // Include the TinyGPS Library
#define gps_reset_pin  4  //GPS Reset control

SoftwareSerial GPS(gps_rxPin, gps_txPin);  // set up a new serial port

TinyGPS gps;  // set up a new GPS object
```

Next, we'll declare the `gpsdump()`, `feedgps()`, and `getgps()` TinyGPS.h library functions. They are declared here so the `setup()` loop can find and use these functions.

```
void gpsdump(TinyGPS &gps);   // Define the gpsdump function
bool feedgps();   // Define the feedgps function
void getGPS();   // define the getGPS function
```

And finally, we define all the variables needed in the sketch:

```
long lat, lon;  // long integer for latitude and longitude function
float LAT, LON;  // floating integer for latitude and longitude values
int year;   // Variable to hold the year value
int gps_altitude;    // Variable to hold the altitude value
int gps_tick = 0, gps_timeout = 120;   // GPS startup timer variables
byte month, day, hour, minute, second;   // variables to hold date and time
unsigned long fix_age, time, date, chars;
String s_date, s_time, s_month, s_year, s_day, s_hour, s_minute, s_second,

grid_text; // String variables for date, time and grid square

// Variable arrays of characters and numbers for the Grid Square function
char A_Z[27] = "ABCDEFGHIJKLMNOPQRSTUVWXYZ";
a_z[27] = "abcdefghijklmnopqrstuvwxyz";
grid[7];

boolean gps_ok;  // gps data ok flag
```

In this sketch we have a lot going on in the `setup()` loop. First, we'll set up the Arduino pins used by the Nokia LCD, Emic 2, and the GPS. We'll also set up the input pin for the `Audio Enable` switch and turn on its internal 20 kΩ pull-up resistor.

```
// set up the serial I/O pins for the Text to Speech module
pinMode(emic_rxPin, INPUT);
pinMode(emic_txPin, OUTPUT);

// set up the serial I/O pins for the GPS module
pinMode(gps_rxPin, INPUT);
pinMode(gps_txPin, OUTPUT);

// Setup the Audio enable pin as an input
pinMode(audio_on, INPUT);
digitalWrite(audio_on, HIGH);   // enable pullup resistor
```

Now, we'll initialize the Nokia LCD and the `Software Serial` ports for the Emic 2 and the GPS:

```
emicSerial.begin(9600);  // set the data rate for the SoftwareSerial port
GPS.begin(9600);   // set the data rate for the SoftwareSerial port

glcd.InitLCD();   // Initialize the Nokia 5110 Display
glcd.setFont(SmallFont);   // Use the small font

  // Display the Startup screen
  glcd.clrScr();
  glcd.print("KW5GP", CENTER,0);
  glcd.print("UTC/GPS", CENTER,8);
  glcd.print("Grid Square", CENTER,16);
  glcd.print("Display", CENTER, 24);
```

Next, we'll wait for the GPS to acquire the satellites. The sketch will wait for up to 2 minutes to acquire a location fix. From a cold start, the SKM53 data sheet says that the maximum time for a satellite fix from a cold start takes a maximum of 36 seconds. Indoors in my lab, I have seen the satellite fix take up to a minute and a half, so the timeout is set for 2 minutes. From a warm restart, the SKM53 will usually acquire the GPS satellites in 30 seconds or less. The LCD will display a message indicating that we are acquiring the satellites and show the timeout timer as it counts down.

```
// Clear the LCD display and indicate we are acquiring satellites
glcd.clrScr();
glcd.print("GPS",CENTER,0);
glcd.print("Acquiring Sats",CENTER,8);
glcd.print("Please Wait",CENTER,32);

// retrieves +/- lat/long in 100000ths of a degree
gps.get_position(&lat, &lon, &fix_age);   // Read the GPS
gps_ok = false;
gps_tick = 0;
// Loop until we start getting valid GPS messages
while (fix_age == TinyGPS::GPS_INVALID_AGE & gps_tick < gps_timeout)
{
  gps.get_position(&lat, &lon, &fix_age);   // Read the GPS
  getGPS();
  glcd.print("No Sat. Fix", CENTER,16);
  // Display the timeout timer
  glcd.print((" "+ String(gps_timeout - gps_tick) + " "),CENTER,40);
  delay(1000);
  gps_tick++;   // Wait a second and decrement the timeout timer
}

if (gps_tick < gps_timeout)   // Check to see if valid message
{
  // We got valid data before timeout, flag the GPS data as valid
  gps_ok = true;
}
```

In the last portion of the `setup()` loop, we'll configure and start up the Emic 2. Remember, because we're running two Software Serial ports, we're not checking the Emic 2 status, assuming everything is okay and it is accepting data, but we'll add a half-second delay when we send data to the Emic 2 to ensure it has time to process the commands.

```
emicSerial.print('\n');       // Send a CR in case the system is already up
delay(500);   // Short delay
emicSerial.println("n6");     // Set voice to Voice 6
delay(500);
emicSerial.print('v');        // Set the volume to +18db
emicSerial.println("18");
delay(500);
emicSerial.print("W");        // Set the speech speed to 200wpm
emicSerial.println("200");
emicSerial.flush();                      // Flush the receive buffer
```

In the main `loop()`, we start off by checking to see if we have valid GPS data. If we do, we set up the display template on the Nokia display and read the GPS.

```
if (gps_ok)   // If we have valid GPS data
{
  glcd.print("Lat:",0,0);   // Set up the Nokia Display with our data template
  glcd.print("Lon:",0,8);
  glcd.print("Sats:",0,16);
  glcd.print("Date:",0,24);
  glcd.print("Time:",0,32);

  getGPS();   // Read the GPS

  // Read GPS Latitude and Longitude into lon, lat and fix_age variables
  gps.get_position(&lat, &lon, &fix_age);

  // Read the GPS Data and Time into data, time and fix_age variables
  gps.get_datetime(&date, &time, &fix_age);
```

Once we have the GPS data, we'll convert the date and time into strings so we can format them for the LCD display:

```
    s_date = String(date);   // convert the date and time to strings
    s_time = String(time);
    if (s_time.length() == 7)
    {
      s_time = "0" + s_time;
    }

    // Break out the date string into Day/Month/Year
    s_year = s_date.substring(4,6);
    s_month = s_date.substring(2,4);
    s_day = s_date.substring(0,2);

    // Break out the time string into Hour/Minute/Second
    s_hour = s_time.substring(0,2);
    s_minute = s_time.substring(2,4);
    s_second = s_time.substring(4,6);
```

Next we'll display the latitude, longitude, the number of satellites used to calculate the position fix, along with the date and time on the LCD. The LCD5110_Basic.h library has some handy formatting functions when displaying floating numbers. Here, the glcd.printNumF() library function is used to select the length, number of decimal places, and the separator character when it outputs floating type variable data.

```
// Display the Latitude and North/South
glcd.printNumF(fabs(LAT/1000000),4,30,0,0x2e,7);
if (LAT > 0)
{
  glcd.print("N",78,0);
} else if (LAT < 0)
{
  glcd.print("S",78,0);
}

// Display the Longitude and East/West
glcd.printNumF(fabs(LON/1000000),4,30,8,0x2e,7);
if (LON > 0)
{
  glcd.print("E",78,8);
} else if (LON < 0)
{
  glcd.print("W",78,8);
}

// Display the number of Satellites we're receiving
glcd.printNumI(int(gps.satellites()),36,16,2);

// Display the Date
glcd.print(s_month + "/" + s_day + "/" + s_year,36,24);
```

```
// Display the Time
glcd.print(s_hour + ":" + s_minute + ":" + s_second,36,32);
```

Next, we'll use a function to calculate the Maidenhead grid locator from the latitude and longitude, and display it along with our altitude. The SKM53 can determine altitude up to 18,000 meters, so even if you do your next Summit On The Air (SOTA) operation from the Himalayas, you'll still have an accurate altitude reading.

```
GridSquare(LAT,LON);   // Calulate the Grid Square
glcd.print(grid_text,0,40);   // Display the Grid Square

gps_altitude = int(gps.f_altitude());   // Read the GPS altitude
glcd.print(String(gps_altitude) + "m ",50,40);   // Display the altitude
```

Finally, we check to see if the Audio Enable switch is turned on and speak the time and grid square if it is enabled:

```
if (digitalRead(audio_on) == LOW)   // Only speak if Audio is enabled
{
  emicSerial.print("S");   // Say the Time
  emicSerial.print("U T C Time is:");
  emicSerial.print(s_hour + " " + s_minute);

  // Say the Grid Square one letter at a time
  emicSerial.print(" Grid Square is:");
  for (int x = 0; x <= 5; x++)
  {
    emicSerial.print(grid_text[x]);
    emicSerial.print(" ");
  }
  emicSerial.print('\n');   // Send a carriage return to start speaking
}
```

If we did not acquire valid GPS data, we'll display that on the LCD and also speak it if the audio output is enabled:

```
// We did not get valid GPS before the startup timed out
glcd.clrScr();
glcd.print("NO GPS Data",CENTER,40);   // Display No GPS Data message
if (digitalRead(audio_on) == LOW)   // Only speak if Audio is enabled
{
  emicSerial.print("S");   // Say the Time
  emicSerial.print("U T C Time is Unknown");
  delay(1000);
  emicSerial.print("Grid Square is Unknown");
  emicSerial.print('\n');   // Send a carriage return to start speaking
}
```

There are three interrelated functions the `TinyGPS.h` library uses to acquire and parse the GPS data, `getGPS()`, `feedgps()`, and `gpsdump()`. The `getGPS()` function checks the GPS data stream every second. If it is new data, it uses the `feedgps()` function to build the NMEA sentence used to extract the GPS information for the sketch. The `gpsdump()` function updates the latitude and longitude variables when new GPS data is received.

```
void getGPS()   // Function to get new GPS data every second
{
  bool newdata = false;
  unsigned long start = millis();
  // Every 1 second we print an update
  while (millis() - start < 1000)
  {
    if (feedgps ())
    {
      newdata = true;
    }
  }
  if (newdata)
  {
    gpsdump(gps);
  }
}

bool feedgps()   // Read the GPS data
{
  while (GPS.available())
  {
    if (gps.encode(GPS.read()))
    return true;
  }
  return 0;
}

void gpsdump(TinyGPS &gps)   //  Get the Latitude and Longitude
{
  gps.get_position(&lat, &lon);
  LAT = lat;
  LON = lon;
  {
  // If we don't feed the GPS during this long routine,
  // we may drop characters and get checksum errors
    feedgps();
  }
}
```

Lastly, we have the `GridSquare()` function used to calculate our Maidenhead grid locator:

```
// Calculate the Grid Square from the latitude and longitude
void GridSquare(float latitude,float longtitude)
{
  // Maidenhead Grid Square Calculation
  // Set up the function variables
  float lat_0,lat_1, long_0, long_1, lat_2, long_2, lat_3, long_3,calc_long;
  float calc_lat, calc_long_2, calc_lat_2, calc_long_3, calc_lat_3;
  lat_0 = latitude/1000000;
  long_0 = longtitude/1000000;
  grid_text = " ";

  int grid_long_1, grid_lat_1, grid_long_2, grid_lat_2;
  int grid_long_3, grid_lat_3;

  // Begin calculating the Grid Square

  // Calculate the first 2 characters of the Grid Square

  // move the longitude reference point to the anti-meridian
  calc_long = (long_0 + 180);

  // move the latitude reference point to the South Pole
  calc_lat = (lat_0 + 90);

  long_1 = calc_long/20; // Break the longitude into 20 degree segments
  lat_1 = (lat_0 + 90)/10; // Break the latitude down into 10 degree segments

  // Remove the fractional portion of the results
  grid_lat_1 = int(lat_1);
  grid_long_1 = int(long_1);

  // Calculate the next 2 digits of the Grid Square
  // Break the Field into 20 degree by 10 degree segments
  calc_long_2 = (long_0+180) - (grid_long_1 * 20);
  long_2 = calc_long_2 / 2;
  lat_2 = (lat_0 + 90) - (grid_lat_1 * 10);

  // Remove the fractional portion of the results
  grid_long_2 = int(long_2);
  grid_lat_2 = int(lat_2);

  // Calculate the last 2 characters of the Grid Square
  calc_long_3 = calc_long_2 - (grid_long_2 * 2);
  long_3 = calc_long_3 / .083333;
  grid_long_3 = int(long_3);

  lat_3 = (lat_2 - int(lat_2)) / .0416665;
  grid_lat_3 = int(lat_3);
```

```
// Here's the first 2 characters of Grid Square - place into array
// We use the array index to the A_Z array to insert the ASCII value
// into the grid[] array
grid[0] = A_Z[grid_long_1];
grid[1] = A_Z[grid_lat_1];

// The second set of the grid square
grid[2] = (grid_long_2 + 48);
grid[3] = (grid_lat_2 + 48);

// The final 2 characters
// We use the array index to the a_z array to insert the ASCII value
// into the grid[] array

grid[4] = a_z[grid_long_3];
grid[5] = a_z[grid_lat_3];

// return the 6 character grid[] array as a String variable
grid_text = grid;

return;

}
```

Once the project was fully debugged and operational on the breadboard, the protoshield, consisting mainly of connectors for the Nokia display, Emic 2 and GPS modules, `Audio Enable` switch, and speaker connections was constructed using the schematic in **Figure 16.6**, and the finished assembly mounted into a Solarbotics Mega S.A.F.E. enclosure.

Enhancement Ideas

This is one project I would really love to downsize with an Arduino Nano to fit inside an Altoids mint tin, turning it into a true shirt-pocket GPS and grid square display. It will be a tight fit, but I think it can be done. You could also upgrade to the `TinyGPS++` library, which allows you to extract data from anywhere within the NMEA sentences, allowing you much more flexibility with the data you have available to display. You could also add code to provide course and velocity output, as well as adding in "course to" code so you can use the GPS as a navigational aid. This project turned out to be a lot of fun to design and construct, and I can see a great many uses for it in the future.

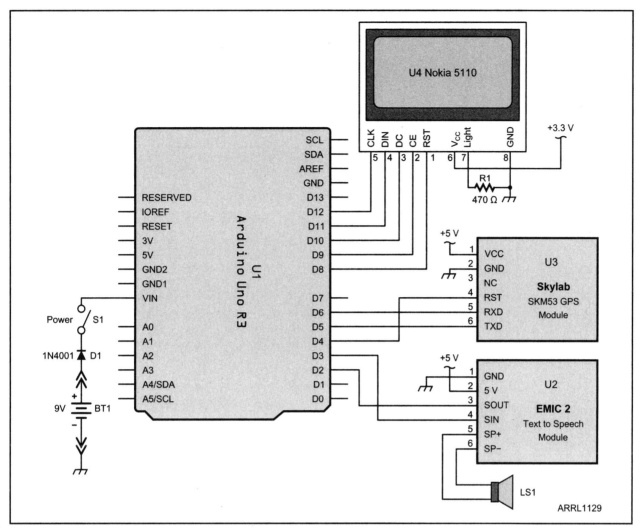

Figure 16.6 — Talking Grid Square Display schematic diagram.

BT1 — 9 V battery.
D1 — 1N4001 diode or equivalent.
LS1 — 8 Ω mini speaker.
R1 — 470 Ω, ⅛ W resistor.
S1 — SPST toggle switch.
U1 — Arduino Uno.
U2 — Emic 2 text-to-speech module.
U3 — Skylab SKM53 GPS module.
U4 — Nokia 5110 LCD display module.

References

American Radio Relay League — **www.arrl.org/digital-modes**
Arduiniana — **www.arduiniana.org**
Arduino — **www.arduino.cc**
Grand Idea Studio — **www.grandideastudio.com**
Henning Karlsen — **www.henningkarlsen.com/electronics/**
Parallax — **www.parallax.com**
Skylab Technology Company — **www.skylab.com.cn**
SparkFun Electronics — **www.sparkfun.com**
Solarbotics — **www.solarbotics.com**
Summits on the Air — **www.sota.org.uk**
TEN-TEC — **www.tentec.com**

CHAPTER 17

Iambic Keyer

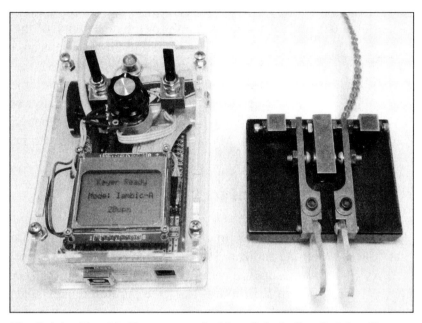

The finished Iambic Keyer mounted in a Solarbotics Arduino Mega S.A.F.E.

It really can't be said enough: the Open Source world is a wonderful place. It's like having a whole Internet full of developers working side by side with you and sharing everything they do. When you start out working with a new module or project, take some time to look for similar projects and ideas on the Internet. While you may not find exactly what you have in mind for your project, there's a good chance you can find someone who has been where you are now and has a similar project you can adapt or that can help you on your way.

Such is the case with the Iambic Keyer project. Electronic CW keyers have been around for decades, using technologies ranging from vacuum tubes to microprocessors. I remember way back when the Curtis keyer chips were hot off the press, and I built my first electronic keyer with the Curtis 8044 chip in 1977. Naturally, I had to build an electronic keyer project for the Arduino. Having been away from the CW world for a while, before I started out with this project, I did some searching on the Internet to see how far keyers have come since my last keyer project.

Iambic Keying

Iambic keying — where the keyer sends alternating dits and dahs when both paddle levers are squeezed, starting with whichever paddle was squeezed first — was a new concept to me when I built my first keyer way back then. Now I've discovered there are not one, but two iambic modes these days. Naturally, this project would have to support both.

Iambic Mode A is derived from the original Curtis Keyer chips. In Iambic Mode A, alternating dits and dahs are sent as long as both paddles are squeezed. When the paddles are released, the keyer completes sending the current dit or dah and then stops, waiting for the next swipe of the paddles.

In Iambic Mode B, alternating dits and dahs are sent as long as both paddles are squeezed, but when the paddles are released, the keyer will complete the current dit or dah, then send the opposite code element before it stops sending.

The choice of iambic modes is purely a matter of personal preference and largely depends on which mode you learned when you got started with an electronic keyer. My advice is to experiment with both modes to see which one you like best.

Keyer Design

Now that I had a good idea of what I was going to do, the next step was to figure out how to do it. Turning to the Open Source community, I did some searching and came across a nice little keyer project for the Arduino that, with minor adaptation, would be perfect for the task. "The Arduino Iambic Keyer" by Steven T. Elliott, K1EL (see the References) had almost everything I wanted in my Iambic Keyer project for the Arduino; it just needed a few things added to make it into exactly what I had in mind. Rather than reinvent the whole wheel, now all I have to do is add my features to the nice foundation Steve has provided. This is the power of the Open Source community and I thank Steve for allowing me to adapt his work into this project.

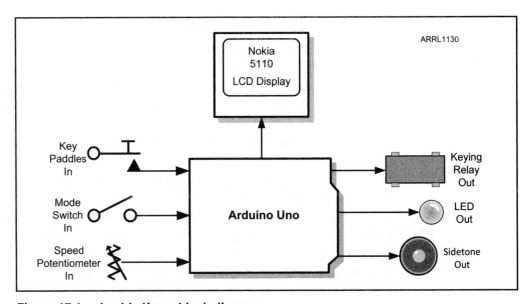

Figure 17.1 — Iambic Keyer block diagram.

Starting out, we'll create our block diagram (**Figure 17.1**) to determine the basic functions in our keyer. We'll be using a SPDT center-off switch to choose the keyer mode. In one direction the switch will select Iambic Mode A, and in the other direction it will select Iambic Mode B. The center-off position will allow us to use either paddle as a straight key. We'll also add a potentiometer to adjust the keying speed. For this project, we'll have the keying speed variable between 5 and 35 WPM. For outputs, we'll have our keyer send a tone to the speaker or key a relay to drive a rig. We'll also have an LED to indicate when the keyer is sending. Lastly, we'll use a Nokia 5110 LCD to display mode and speed settings.

As you can see from the Fritzing diagram in **Figure 17.2**, we have quite a few external components on this project, and we'll be using most of the Arduino Uno's digital I/O pins, as well as one of the analog pins. Even with the number of external components, wiring this project up on the breadboard should not take much time at all.

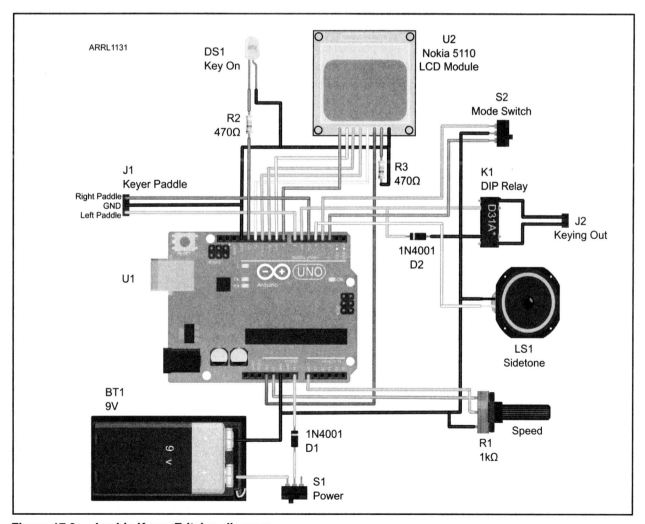

Figure 17.2 — Iambic Keyer Fritzing diagram.

The Sketch

Once you have everything in place, we're ready to start creating the sketch. Using the flowchart in **Figure 17.3** you can see that the coding required for this project will be slightly more involved that most of the coding we have done to this point. Using Steve Elliott's Iambic Keyer sketch as the core for this project, all we have to do is add in our external components, so it's really not as difficult as it looks.

Starting out with the sketch, we'll include the library for the Nokia 5110 display and define our constants and variables. The complete sketch can be found in **Appendix A** and online at **www.w5obm.us/Arduino**.

```
#include <LCD5110_Basic.h>    // Nokia 5110 LCD

LCD5110 glcd(12,11,10,8,9);
extern uint8_t SmallFont[];

#define ST_Pin 4          // Sidetone Output Pin on Pin 4
#define LP_in 7           // Left Paddle Input on Pin 7
#define RP_in 5           // Right Paddle Input on Pin 5
#define led_Pin 13        // LED on Pin 13
#define Mode_A_Pin 3      // Mode Select Switch Side A
#define Mode_B_Pin 2      // Mode Select Switch Side B
#define key_Pin 6         // Transmitter Relay Key Pin
#define Speed_Pin 0       // Speed Control Pot on A0
#define ST_Freq 600       // Set the Sidetone Frequency to 600 Hz

int key_speed;            // variable for keying speed
int read_speed;           // variable for speed pot
int key_mode;             // variable for keying mode
int last_mode;            // variable to detect keying mode change

unsigned long ditTime;   // Number of milliseconds per dit
char keyerControl;
char keyerState;
String text1, text2, text3, text4, text5, text6 = " ";  // LCD text variables

static long ktimer;
int debounce;
```

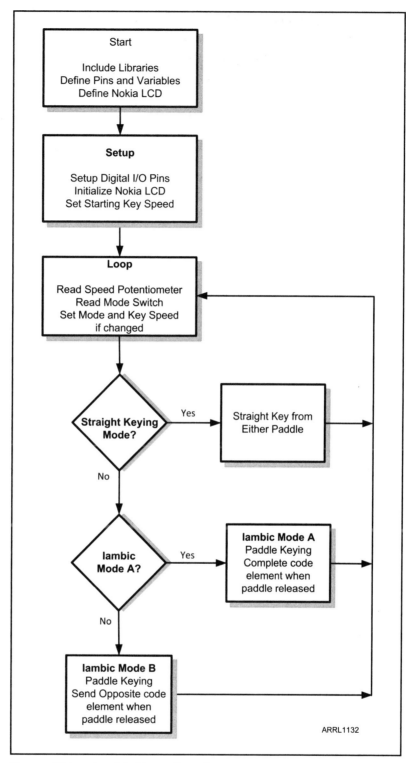

Figure 17.3 — Iambic Keyer flowchart.

The Iambic Keyer primary functions are all managed and controlled through the use of a single byte variable to keep track all the various conditions. A single bit is used, rather than an entire integer variable for each condition. This is an excellent way to conserve memory, especially with sketches that push memory usage to the limits and you're looking for every last byte of memory you can possibly conserve.

```
//      keyerControl bit definitions
//
#define     DIT_L       0x01        // Dit latch
#define     DAH_L       0x02        // Dah latch
#define     DIT_PROC    0x04        // Dit is being processed
#define     PDLSWAP     0x08        // 0 for normal, 1 for swap
#define     IAMBICB     0x10        // 0 for Iambic A, 1 for Iambic B
#define     ULTIMATIC   0x20        // 1 for ultimatic
#define     STRAIGHT    0x80        // 1 for straight key mode
```

In his Iambic Keyer sketch, Steve Elliott introduces the `enum` datatype, which allows you to use enumerated values to represent the various conditions the variable can contain. This datatype allows you to assign an easily understandable alias to a condition that can be used in other Arduino statements, such as within `switch...case()` decision blocks within your sketch. For example, it's a whole lot easier to read and understand `case IDLE...do this,` than it is to understand `case 1...do this`.

```
//  State Machine Defines

enum KSTYPE {IDLE, CHK_DIT, CHK_DAH, KEYED_PREP, KEYED, INTER_ELEMENT };
```

In the `setup()` loop, we start out by defining all of the digital I/O pin modes we'll need to use to interface with all the external pieces we have on the breadboard. You will see how much more readable all this setup is by defining the pins with an understandable name. This also allows you to easily switch your pins around if you would prefer to connect the external components to different pins, or you make a wiring error and would rather change the pin assignments than rewire your circuit.

```
pinMode(led_Pin, OUTPUT);           // sets the LED digital pin as output
pinMode(LP_in, INPUT);              // sets Left Paddle digital pin as input
pinMode(RP_in, INPUT);              // sets Right Paddle digital pin as input
pinMode(ST_Pin, OUTPUT);            // Sets the Sidetone digital pin as output
pinMode(key_Pin, OUTPUT);           // Sets the Keying Relay pin as output
pinMode(Mode_A_Pin, INPUT);         // sets Iambic Mode Switch Side A as input
pinMode(Mode_B_Pin, INPUT);         // sets Iambic Mode Switch Side B as input
digitalWrite(led_Pin, LOW);         // turn the LED off
digitalWrite(LP_in, HIGH);          // Enable pullup on Left Paddle Input Pin
digitalWrite(RP_in, HIGH);          // Enable pullup on Right Paddle Input Pin
digitalWrite(Mode_A_Pin, HIGH);     // Enable pullup on Mode Switch Side A
digitalWrite(Mode_B_Pin, HIGH);     // Enable pullup on Mode Switch Side B
```

Next, we'll initialize the Nokia 5110 LCD and output a startup message. With this sketch, you'll see that we have assigned a string variable for each line of text on the LCD display and use a function to update the display. This will allow us to update multiple lines on the display simultaneously without having to print to every line of the LCD every time we want to update the display.

```
glcd.InitLCD(60);
glcd.setFont(SmallFont);

cleartext();
text1 = "KW5GP";
text2 = "Iambic";
text3 = "Keyer";
text6 = "Initializing";
updatelcd();
delay(3000);
```

In the next part of the `setup()` loop, we initialize the keyer state machine and the control byte. We also read the SPEED potentiometer and map the analog voltage input from the SPEED potentiometer to a value between 5 and 35 WPM. We then call the `loadWPM()` function to set the keyer speed to the value read from the SPEED potentiometer.

```
keyerState = IDLE;
keyerControl = 0;
// Read the potentiometer to determine code speed
key_speed = map(analogRead(Speed_Pin),10,1000,5,35);
loadWPM(key_speed);    // Set the keying speed
```

In the final section of the `setup()` loop, we clear and display the starting status of the keyer on the Nokia display:

```
cleartext();
text1 = "Keyer Ready";
text3 = "Mode: Iambic-A";
text5 = String(key_speed) + "wpm";
updatelcd();
```

In the main `loop()`, the first thing we do is read the current setting on the SPEED potentiometer and update the keyer sending speed and the speed display on the LCD if the speed has changed.

```
// Read the potentiometer to determine code speed
read_speed = map(analogRead(Speed_Pin),10,1000,5,35);
// If the Speed Pot has changed, update the speed and LCD
if (key_speed != read_speed)
{
  key_speed = read_speed;
  loadWPM(key_speed);
  text5 = String(key_speed) + "wpm";
  updatelcd();
}
```

Next we check to see if the keying mode has been changed and update the LCD display only if the mode has changed:

```
// Read the Mode Switch and set mode
//
// Key Mode 0 = Iambic Mode A
// Key Mode 1 = Iambic Mode B
// Key Mode 2 = Straight Key
//

if (digitalRead(Mode_A_Pin) == LOW)    // Set Iambic Mode A
{
  key_mode = 0;
  text3 = "Mode: Iambic-A";
}

if (digitalRead(Mode_B_Pin) == LOW)    // Set Iambic Mode B
{
  key_mode = 1;
  text3 = "Mode: Iambic-B";
}

// Set Straight Key if Mode switch in Center-Off position
if (digitalRead(Mode_A_Pin) == HIGH && digitalRead(Mode_B_Pin) == HIGH)
{
  key_mode = 2;
  text3 = "Mode: Straight";
}

// Update the LCD and save the new mode only if it has changed
if (key_mode != last_mode)
{
  last_mode = key_mode;
  updatelcd();
}
```

Next, we check to see if we are in straight key mode or iambic keying mode. If we're in straight key mode, we just use either paddle as a keying input, turn on the LED, key the keying relay, and play the sidetone as the paddles are pressed.

```
if (key_mode == 2) // Straight Key
{
  // Straight Key Mode
  if ((digitalRead(LP_in) == LOW) || (digitalRead(RP_in) == LOW))
  {
    // Key from either paddle
    digitalWrite(led_Pin, HIGH);
    digitalWrite(key_Pin, HIGH);
    tone(ST_Pin, 600);
  } else {
    digitalWrite(led_Pin, LOW);
    digitalWrite(key_Pin, LOW);
    noTone(ST_Pin);
  }
```

Here is where Steve Elliott's keyer code comes into play. Using a `switch...case()` statement, we check the various conditions of the state machine and set the `keyerControl` variable's bits accordingly.

```
// Basic Iambic Keyer
  // keyerControl contains processing flags and keyer mode bits
  // Supports Iambic A and B
  // State machine based, uses calls to millis() for timing.
  switch (keyerState)
  {
    case IDLE:      // Wait for direct or latched paddle press
    // check to see if a paddle is pressed or keyControl bits DIT_L or
    // DAH_l (dit or dah latch) are set
      if ((digitalRead(LP_in) == LOW) || (digitalRead(RP_in) == LOW) ||
(keyerControl & 0x03))
      {
        update_PaddleLatch(); // set the paddle latch bits DIT_L or DAH_L
        keyerState = CHK_DIT;  // Tell state machine to check for a dit next
      }
      break; // Exit the case() loop
```

```
case CHK_DIT:       // See if the dit paddle was pressed
  if (keyerControl & DIT_L)
  {
    keyerControl |= DIT_PROC; // set the DIT_PROC bit in keyerControl
    ktimer = ditTime; // set the keying timer to the ditTime
    keyerState = KEYED_PREP; // tell the state machine we're ready to key
  } else {
    keyerState = CHK_DAH; // otherwise, do a CHK_DAH next time
  }
  break;

case CHK_DAH:       // See if dah paddle was pressed
  if (keyerControl & DAH_L)
  {
    ktimer = ditTime*3;   // set the key time to 3 times the ditTime
    keyerState = KEYED_PREP; // tell the state machine we're ready to key
  } else {
    keyerState = IDLE; // no key is pressed, return to IDLE state
  }
  break;
```

Next we check for the keying portions of the state machine:

```
case KEYED_PREP:
 // Assert key down, start timing, state shared for dit or dah
    digitalWrite(led_Pin, HIGH);        // Turn the LED on
    tone(ST_Pin, ST_Freq);              // Turn the Sidetone on
    digitalWrite(key_Pin, HIGH);        // Key the TX Relay
    ktimer += millis();                 // set ktimer to interval end time
    keyerControl &= ~(DIT_L + DAH_L);   // clear both paddle latch bits
    keyerState = KEYED;                 // next state
    break;

case KEYED:        // Wait for timer to expire
  if (millis() > ktimer)   // are we at end of key down ?
  {
    digitalWrite(led_Pin, LOW);         // Turn the LED off
    noTone(ST_Pin);                     // Turn the Sidetone off
    digitalWrite(key_Pin, LOW);         // Turn the TX Relay off
    ktimer = millis() + ditTime;        // inter-element time
    keyerState = INTER_ELEMENT;         // next state
  }
  else if (key_mode == 1) // Check to see if we're in Iambic B Mode
  {
    update_PaddleLatch();  // early paddle latch in Iambic B mode
  }
  break;
```

The last state of the state machine is to set the inter-element spacing between the code elements:

```
case INTER_ELEMENT:        // Insert time between dits/dahs
    update_PaddleLatch();       // latch paddle state
    if (millis() > ktimer)  // are we at end of inter-space ?
    {
      if (keyerControl & DIT_PROC)      // was it a dit or dah ?
      {
        keyerControl &= ~(DIT_L + DIT_PROC);    // clear two bits
        keyerState = CHK_DAH;                   // dit done, check for dah
      } else {
        keyerControl &= ~(DAH_L);               // clear dah latch
        keyerState = IDLE;                      // go idle
      }
    }
    break;
```

The Functions

There are four functions used in this sketch. The first, update_PaddleLatch(), is used to update the status of the keyerControl byte when a paddle is pressed:

```
void update_PaddleLatch()
{
    if (digitalRead(RP_in) == LOW) {  // set the DIT_L bit if right paddle
        keyerControl |= DIT_L;
    }
    if (digitalRead(LP_in) == LOW) {  // set the DAH_L bit if left paddle
        keyerControl |= DAH_L;
    }
}
```

The loadWPM() function calculates the desired dit timing from the selected words per minute:

```
void loadWPM (int wpm)
{
    ditTime = 1200/wpm;
}
```

The last two functions handle the output to the Nokia 5110 LCD. The `updatelcd()` function will take the contents of the six text variables and output them on each line of the Nokia display. The `cleartext()` function will erase the display and reset the text variables.

```
void updatelcd()      // clears LCD display and writes the LCD data
{
  glcd.clrScr();
  glcd.print(text1,CENTER,0);   // Line 0
  glcd.print(text2,CENTER,8);   // Line 1
  glcd.print(text3,CENTER,16);  // Line 2
  glcd.print(text4,CENTER,24);  // Line 3
  glcd.print(text5,CENTER,32);  // Line 4
  glcd.print(text6,CENTER,40);  // Line 5
}

void cleartext()   // clears the text data
{
  text1 = " ";
  text2 = text1;
  text3 = text1;
  text4 = text1;
  text3 = text1;
  text4 = text1;
  text5 = text1;
  text6 = text1;
}
```

As you can see, there is a lot going on inside the iambic keyer portion of the sketch, but because Steve had already done this part of the sketch, we saved a lot of time and effort by utilizing his sketch as a foundation and wrapping our modifications around it. This is part of the fun of the Arduino and Open Source; you're rarely totally alone and as this sketch proves, there's most likely someone out there doing something similar to what you're wanting to do and sharing their work to help you with your project.

Once you have everything working on the breadboard, you can use the schematic in **Figure 17.4** to build your keyer on an Arduino protoshield and mount the finished project into a Solarbotics Mega S.A.F.E. enclosure.

Enhancement Ideas

Between Steve Elliott's keyer sketch and my enhancements, there's really not much more I can think of adding to this project. The one idea that does come to mind is to shrink the project down using an Arduino Nano, rearranging the controls, fitting everything into a single-height enclosure, and mounting your keyer paddles on top, turning it into a self-contained portable keyer for that next Field Day or other portable operating event.

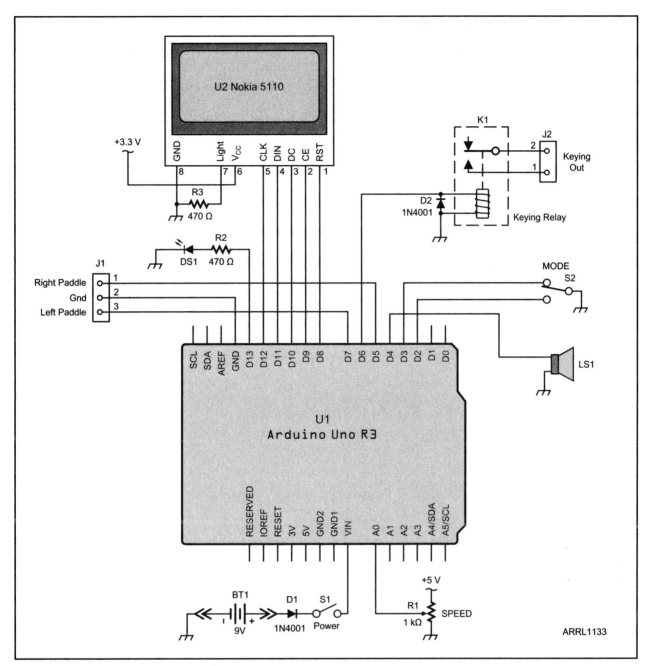

Figure 17.4 — Iambic Keyer schematic diagram.

BT1 — 9 V battery.
D1, D2 — 1N4001 diode or equivalent.
DS1 — 5 mm LED.
J1 — Three conductor jumper header or builder's choice of jack.
J2 — Two conductor jumper header or builder's choice of jack.
K1 — SPST 5 V DIP reed relay.

LS1 — 8 Ω mini speaker.
R1 — 1 kΩ potentiometer.
R2, R3 — 470 Ω, ⅛ W resistor.
S1 — SPST toggle switch.
S2 — DPST center-off toggle switch.
U1 — Arduino Uno.
U2 — Nokia 5110 LCD display module.

References

Henning Karlsen — **www.henningkarlsen.com/electronics/**
Solarbotics — **www.solarbotics.com**
S. Elliott, K1EL, "Arduino Iambic Keyer," **www.openqrp.org/?p=343**

CHAPTER 18

Waveform Generator

The finished Waveform Generator mounted in a SparkFun pcDuino/Arduino enclosure.

A waveform generator is another tool that's handy to have around. If you're going to troubleshoot or work with audio circuits, it's always nice to have a stable and clean audio source available. Since I can't whistle, I'm pretty much out of luck when it comes to testing audio circuits without something to generate the audio for me.

This is one of those projects that kept evolving as I worked with it on the breadboard. Had I just soldered up the original design, I would have spent more time soldering (and desoldering) than I would have spent actually building the project. This is a perfect example of why building and testing your initial designs on a breadboard is the preferred way to go.

My initial design criteria called for a waveform generator that could generate a sine, square, and triangle wave from 1 Hz to 10 kHz or better. Since the Arduino can only generate square waves on the digital I/O pins, and has no digital-to-analog converter onboard, I would have to come up with a way to generate the waveforms externally. That sounded simple enough in theory, as I had already found numerous circuits that could do this with the Arduino online in the Arduino Playground.

Resistor-to-Resistor Ladder Network

The simplest design (**Figure 18.1**) called for a series of resistors connected to the Arduino's digital I/O pins that would create what as is known as a "resistor-to-resistor ladder network" (R2R) digital-to-analog (D/A) converter. This design uses a series of resistors connected across eight digital I/O pins on the Arduino to generate the analog output. In effect, this creates an 8-bit digital-to-analog converter, and the Arduino would be used to generate the actual waveforms by using eight digital I/O pins as the input to the R2R D/A network.

The test sketch for this was simple, and there were plenty of example sketches using a table stored in flash memory to generate the sine wave. To generate a triangle wave, all I had to do was increment and decrement the data sent to the digital I/O pins. Sounds easy enough, right?

Well, at 200 Hz, everything was perfect as shown in **Figure 18.2**. However, as the frequency went up, the waveform started to distort. At 6 kHz (**Figure 18.3**), there was a noticeable stair-stepping to the waveform, and at 13 kHz (**Figure 18.4**), the sine wave was seriously distorted. At 27 kHz, the sine wave

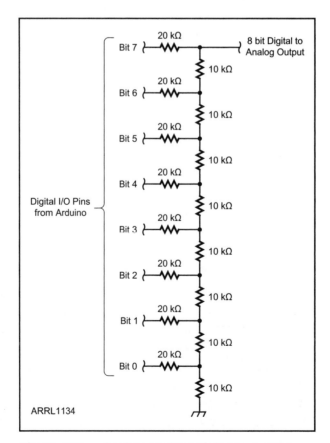

Figure 18.1 — Resistor ladder digital-to-analog converter schematic diagram.

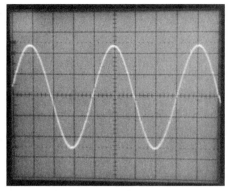

Figure 18.2 — The resistor ladder digital-to-analog converter output at 200 Hz. It doesn't look bad at all here.

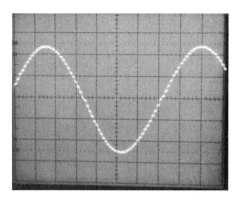

Figure 18.3 — The resistor ladder digital-to-analog converter output at 6 kHz. You can see where the waveform is starting to show stairstepping.

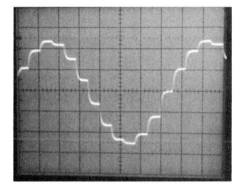

Figure 18.4 — The resistor ladder digital-to-analog converter output at 13 kHz. You can see where the waveform is getting badly distorted with noticeable stairstepping.

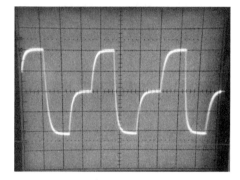

Figure 18.5 — The resistor ladder digital-to-analog converter output at 27 kHz. The waveform is so badly distorted now that we can't really tell if it's a sine or a square wave.

was so distorted that it was more like a bad square wave than a sine wave as seen in **Figure 18.5**. It was the same way with a triangle wave, except the distortion was very noticeable at 8 kHz, and at 12.5 kHz the triangle wave was distorted to where it began to look more like a very bad sine wave than a triangle wave.

I tried multiple versions of the sketch and did all that I could think of to optimize the waveforms, including direct calculation of the sine wave rather than using a pre-defined table. The results were every bit as bad, if not worse. Even using the Arduino's digital I/O port register I/O method that outputs to eight digital I/O pins with a single `digitalWrite()` didn't improve things very much. It was pretty obvious that for any frequency above 800 Hz, this design wasn't going to provide a clean, stable waveform unless I moved up to a faster Arduino such as the Due. I felt that was a bit too big of a step for what should be a simple waveform generator project, so it was time to fall back and regroup.

Microchip MCP4725 I²C 12-bit D/A Module

Figure 18.6 — The Microchip MCP4725 12-bit I²C digital-to-analog converter module.

The next design called for replacing the resistor ladder D/A with a Microchip MCP4725 I²C 12-bit D/A module (**Figure 18.6**). My thinking was that the MCP4725 D/A would provide four additional bits of resolution and would offload some work from the Arduino since the MCP4725 uses the I²C bus for communication. Some quick rewiring on the breadboard and the new design was ready to test. As it turned out, the waveforms were slightly less distorted, but the Arduino and MCP4725 couldn't generate the waveforms fast enough to produce anything usable above 800 Hz. It was fairly obvious that this design wasn't going to cut it either.

Things were not looking good for the Waveform Generator project at this point. I had exhausted all of the traditional ways to generate waveforms with the Arduino, and there was only one design idea left to try.

Analog Devices AD9833 Programmable Waveform Generator

I had recently begun to experiment with the direct digital frequency synthesis (DDS) modules as part of the TEN-TEC Rebel Model 506 JT65 project, as well as working on some ideas for a DDS-based antenna analyzer, so I had a few different DDS modules lying around the lab. I had been avoiding them after skimming the datasheet because it seemed they were quite complex, and I just hadn't taken the time to dig into them yet. Now was as good a time as any, so I dug back into the datasheets to see if I could use a DDS module for this project. I settled on trying to use the Analog Devices AD9833 programmable waveform generator module (**Figure 18.7**). The reason that I chose the AD9833 was that of all the DDS modules available at a reasonable price, the AD9833 was one of the few that would directly generate triangle waves in addition to the standard sine and square wave outputs of the other DDS modules.

Direct digital frequency synthesis is a relatively new technology that is rapidly finding its way into circuit designs that call for generating high-quality varying frequency signals. Using a reference clock, a DDS can be used to generate very precise and stable signals that can be used as a VFO for a transmitter or as a receiver IF mixer signal input, for example.

What makes the DDS modules so appealing is that they use a standard SPI bus interface for communication with the Arduino or other microcontrollers. Controlling the Analog Devices DDS modules uses the same basic internal register format among the various DDS modules, so once you get one DDS design working, you can switch among the various DDS modules rather easily. The major difference among the Analog Devices DDS modules is their upper frequency limit, number of bits in the frequency and phase control registers, and the types of waveforms they can generate.

The TEN-TEC Rebel transceiver uses an AD9834 to generate the transmitted CW signal and the receiver IF mixing signal. The AD9834 can

Figure 18.7 — The Analog Devices AD9833 waveform generator module.

generate clean and stable waveforms up to 37.5 MHz, perfect for use in amateur HF transmitter or receiver designs. The AD9850 can generate signals up to 62.5 MHz and the AD9851 can generate signals up to 90 MHz. However, since the AD9834, AD9850 and AD9851 don't generate triangle waves, and this project's initial design called for generating sine, square, and triangle waves, with no requirement for generating waveforms higher than 20 kHz, the AD9833 seemed to be the best candidate for the job.

The Analog Devices AD9833 is a digitally programmable waveform generator, capable of generating sine, square, and triangle waveforms from 0 to 12.5 MHz. It has dual 28-bit frequency registers and can achieve a resolution of 0.1 Hz with a 25 MHz reference clock rate. Even higher resolutions can be achieved by using a lower frequency for the reference clock. The AD9833 also has two 12-bit phase registers, allowing the phase of the output signal to be shifted from 0 to 720° (0 – 2π). The AD9833 uses a 10-bit D/A converter to generate the output waveform. Sine waves are generated using an on-chip sine-wave lookup table in internal read-only memory (ROM), requiring no external calculations to generate smooth, 10-bit sine waves across the entire frequency range. The output signal from the AD9833 is approximately 0.6 V peak-to-peak.

The AD9833 is controlled through a single 16-bit multi-function control resister. The upper two bits of the control register are used to determine if the command is a control command, frequency register 0 or 1 command, or a phase register command. Since the frequency registers are 28 bits each, two 16-bit writes are required to load the frequency registers. Command register options allow you to control how the frequency registers are loaded, giving you a number of options for frequency control and selection.

For me, the most difficult part of understanding the AD9833 was with the way the various commands are implemented. Once I realized that all communication with the AD9833 is through the single 16-bit control register with different upper bit settings to select the actual register desired, the rest came fairly easily. To calculate the desired output frequency, the simple formula of $f_{out} = (f_{refClk}/2^{28}) \times FreqRegister$ is all that is needed. We can have the Arduino do the complex math for us, so all we have to do is set up a function to convert the desired frequency into the value needed for the frequency register.

Final Design

Finally, we have a circuit that looks like it can do what our design calls for. Using the block diagram (**Figure 18.8**), we'll use an AD9833 to generate the waveforms, an SPDT center-off switch to select among sine, square, and triangle waves, and a 1 kΩ potentiometer to control the output frequency. We'll use an LM386 audio amplifier chip to boost the output of the DDS module to provide an output signal strong enough to drive a speaker. We'll also have a potentiometer on the LM386 to control the output level.

Now that we finally have all the pieces in the right order, we can create our prototype circuit using the Fritzing diagram in **Figure 18.9**. In spite of all the difficulties determining the best way to build this circuit, the actual hardware involves only a handful of wires and external components. We'll have the AD9833 attached to the Arduino's SPI bus and the 16-character by 2-line LCD display connected to the I²C bus.

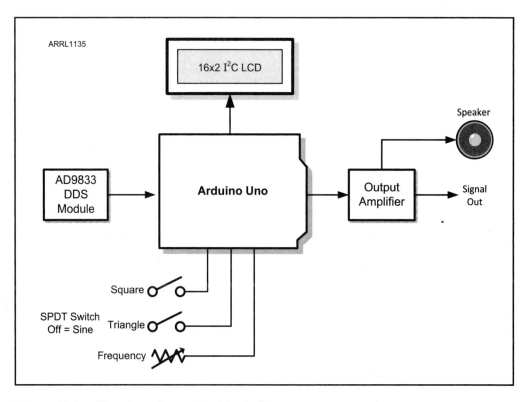

Figure 18.8 — Waveform Generator block diagram.

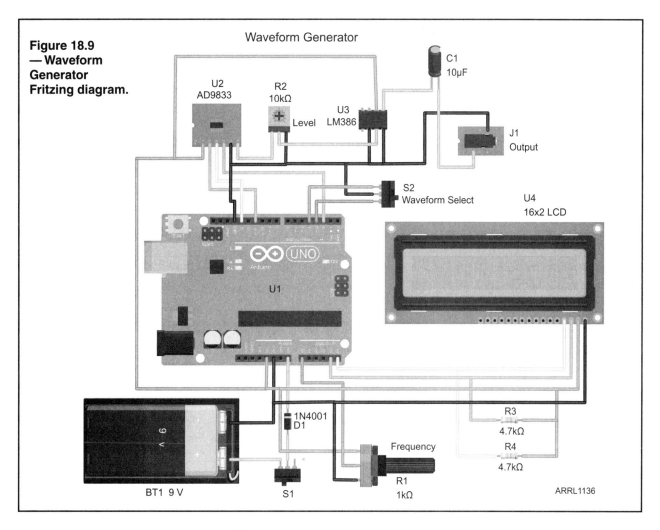

Figure 18.9 — Waveform Generator Fritzing diagram.

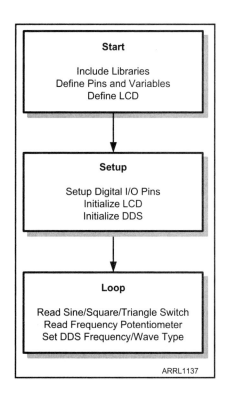

The Sketch

Figure 18.10 shows the flowchart we will use to design our sketch. After initialization and setup, the sketch will read the sine/square/triangle WAVEFORM SELECT switch, read the FREQUENCY potentiometer, and set the output waveform and frequency of the DDS module accordingly.

As I mentioned earlier, most of the Analog Devices DDS chips are interfaced using the same I/O methods and register control bits, so I was able to adapt an existing sketch designed for the AD9837 DDS that I found online to use as the core functions in the Waveform Generator sketch.

Starting out with creating the sketch, we'll need to include the SPI.h library for the AD9833, and the Wire.h and LiquidCrystal_I2C.h libraries for the 16×2 I²C LCD, along with our pin and other definitions. The complete sketch can be found in **Appendix A** and online at **www.w5obm.us/Arduino**.

Figure 18.10 — Waveform Generator flowchart.

```
#include <SPI.h>   // SPI Library
#include <Wire.h>      //I2C Library
#include <LiquidCrystal_I2C.h>  // Liquid Crystal I2C Library

#define FSYNC 2  // Define FSYNC Pin on Digital Pin 2
#define Square_wave 3   // Define Square Wave Switch on Digital Pin 3
#define Triangle_wave 4   // Define Triangle Wave Switch on Digital Pin 4
#define min_freq 50   // Define the minimum frequency as 50Hz
#define max_freq 10000   // Define the maximum frequency as 10 KHz
#define frequency_pot A0   // Define the Frequency Adjust pot on Analog Pin 0
```

Next we'll define the variables we'll need and create the LCD object:

```
long debounce = max_freq/1000;   // debounce the frequency control pot
long freq; //32-bit global frequency variable

// variables for frequency selection
long previous_frequency = 0, desired_frequency ;

// variables for waveform selection
int wave_type, previous_type = 0, wave_data = 0x2000;

const int lcd_end = 16; // set width of LCD
const int lcd_address = 0x27; // I2C LCD Address
const int lcd_lines = 2; // Number of lines on LCD

// set the LCD I2C address to 0x27 for a 16 chars and 2 line display
LiquidCrystal_I2C lcd(lcd_address,lcd_end,lcd_lines);
```

In the setup() loop, the majority of the work is spent setting up the 16×2 LCD and displaying a brief startup message so you know everything is going well to this point.

```
lcd.init();   // initialize the LCD
lcd.backlight();   // Turn on the LCD backlight
lcd.home();   // Set the cursor to line 0, column 0

lcd.print(" KW5GP Waveform");   // Display the startup screen
lcd.setCursor(3,1);
lcd.print("Generator");
delay(3000);
lcd.clear();

lcd.print("Freq: ");   // Set up the LCD display
lcd.setCursor(0,1);
lcd.print("Sine Wave");
```

In the last part of the setup() loop, we define the pin modes for the AD9833 and the WAVEFORM SELECT switch, enable the 20 kΩ internal pull-up resistors on the WAVEFORM SELECT switch pins, initialize the AD9833, and start the SPI bus. Note that the AD9833 requires SPI Mode 2 to communicate via the SPI bus. SPI Mode 2 will clock the data from the Arduino to the AD9833 on the falling edge of the SCK signal. Failure to select SPI Mode 2 will cause communication issues with the AD9833, since the default mode for SPI is Mode 0, where the data is clocked out on the rising edge of the SCK signal.

```
pinMode(FSYNC, OUTPUT); // Set the FSYNC Pin as an Output

// Set the Square_wave Control Pin as an Input
pinMode(Square_wave, INPUT);

// Set the Triangle_wave Control Pin as an Input
pinMode(Triangle_wave, INPUT);

// Enable the Internal Pullup Resistor on the Square wave Control Pin
digitalWrite(Square_wave, HIGH);

// Enable the Internal Pullup resistor on the Triangle wave Control Pin
digitalWrite(Triangle_wave, HIGH);

digitalWrite(FSYNC, HIGH);  // Set FSYNC High - disables input on the AD9833

SPI.setDataMode(SPI_MODE2); // requires SPI Mode 2 for AD9833
SPI.begin();   // Start the SPI bus

delay(100); //A little set up time, just to make sure everything's stable
```

The main `loop()` for the sketch is relatively short and to the point. The majority of the work dealing with the AD9833 is handled by two function calls.

```
wave_type = 0;  // Default to Wave Type 0 - Sine Wave

if (digitalRead(Square_wave) == LOW)  // Check the Square Wave switch pin
{
  wave_type = 1;  // Set the Wave Type to 1 for a Square wave
}

if (digitalRead(Triangle_wave) == LOW)  // Check the Triangle Wave switch pin
{
  wave_type = 2;  // Set the Wave Type to 2 for a Triangle Wave
}

// Read the frequency pot to determine desired frequency
desired_frequency = map(analogRead(frequency_pot),1,1020,min_freq,max_freq);
```

Next, we'll update the AD9833 frequency and mode settings if anything has changed. We "debounce" the FREQUENCY SELECT potentiometer to prevent frequency changes if the analog-to-digital conversion value from the potentiometer changes slightly due to noise. After updating the AD9833, we display any changes to the LCD.

```
// Update the DDS frequency if we've changed frequency or wave type
  if (desired_frequency > (previous_frequency+debounce) || desired_frequency <
(previous_frequency - debounce) || (wave_type != previous_type))
  {
   // Call the Function to change frequency and/or waveform type
    WriteFrequencyAD9833(desired_frequency);
    previous_frequency = desired_frequency;  // Update the frequency variable
    previous_type = wave_type;  // update the wave type variable
    lcd.setCursor(0,1);  // Display the Wave Type on the LCD
    switch(wave_type)
    {
      case 0:   // Type 0 = Sine Wave
        lcd.print("Sine Wave     ");
        break;

      case 1:   // Type 1 = Square Wave
        lcd.print("Square Wave   ");
        break;

      case 2:   // Type 2 = Triangle Wave
        lcd.print("Triangle Wave");
        break;
    }
  }
```

There are two functions used to communicate with the AD9833. The `WriteFrequencyAD9833()` function is used to set the selected frequency and waveform type:

```
// Function to change the frequency and/or waveform type
void WriteFrequencyAD9833(long frequency)
{

  //
  int MSB;   // variable for the upper 14 bits of the frequency
  int LSB;   // variable for the lower 14 bits of the frequency
  int phase = 0;  // variable for phase control

  // We can't just send the actual frequency, we have to calculate the
  // "frequency word". This works out to
  // ((desired frequency)/(reference frequency)) x 0x10000000.
  // calculated_freq_word will hold the calculated result.

  long calculated_freq_word;  // variable to hold calculated frequency word
```

```
  float AD9833Val = 0.00000000;  // variable to calculate frequency  word

  // Divide the desired frequency by the DDS Reference Clock
  AD9833Val = (((float)(frequency))/25000000);
  // Divide the frequency by 2 if we're generating Square waves
  if (wave_type == 1)
  {
    AD9833Val = AD9833Val * 2;
  }

  // Finish calculating the frequency word
  calculated_freq_word = AD9833Val*0x10000000;

// Display the current frequency on the LCD
lcd.setCursor(6,0);
lcd.print(String (frequency) + " Hz   ");

// Once we've got the calculated frequency word, we have to split
// it up into separate bytes.
MSB = (int)((calculated_freq_word & 0xFFFC000)>>14); // Upper 14 bits
LSB = (int)(calculated_freq_word & 0x3FFF);   // Lower 14 bits

// Set control bits DB15 and DB14 to 0 and one, respectively, to
// select frequency register 0
LSB |= 0x4000;
MSB |= 0x4000; // Has to be done for both frequency words

phase &= 0xC000;  // Set the Phase Bits (defaults to 0)

// Set the Control Register to receive frequency LSB and MSB
//in consecutive writes
WriteRegisterAD9833(0x2000);

//Set the frequency
// Write the lower 14 bits to the Frequency Register of the AD9833
WriteRegisterAD9833(LSB); //lower 14 bits

// Write the upper 14 bits to the Frequency Register of the AD9833
WriteRegisterAD9833(MSB); //upper 14 bits

// Write the phase bits to the Phase Register
WriteRegisterAD9833(phase); //mid-low

// Select the correct Register settings for the desired Waveform
switch (wave_type)
{
  case 0:  // Sine Wave
  wave_data = 0x2000;
```

```
      break;

    case 1:   // Square Wave
      wave_data = 0x2020;
      break;

    case 2:   // Triangle Waqve
      wave_data = 0x2002;
      break;
  }
  WriteRegisterAD9833(wave_data);   // Write the Waveform type to the AD9833
}
```

The `WriteRegisterAD9833()` function is the part of the sketch that performs the actual SPI bus transfer of data from the Arduino to the AD9833.

```
// Function to write the data to the AD9833 Registers
void WriteRegisterAD9833(int dat)
{
  //Set FSYNC low - Enables writing to the DDS Registers
  digitalWrite(FSYNC, LOW);

  SPI.transfer(highByte(dat)); // Send the High byte of data
  SPI.transfer(lowByte(dat));  // Send the Low byte of data

  //Set FSYNC high  - Disable writing to the DDS Registers
  digitalWrite(FSYNC, HIGH);
}
```

Once the Waveform Generator was finally up and running on the breadboard, the output amplifier circuit was added. I chose to use a Texas Instruments LM386 audio power amplifier chip. The LM386 is a single power supply voltage, self-contained audio power amplifier, with a selectable gain from 20 to 200, bandwidth of 300 kHz, and an output of 325 mW. I chose the LM386 over a standard op amp because the LM386 does not require a feedback resistor or any other external components to operate at its default gain of 20. The 300 kHz bandwidth of the LM386 was more than adequate for what the design of the Waveform Generator called for. If you desire your Waveform Generator to operate up to the full 12.5 MHz bandwidth of the AD9833 DDS module, you'll need to replace the LM386 circuit with an op amp circuit capable of handling the higher frequencies.

Once the LM386 was added to the breadboard, along with a 10-turn 10 kΩ potentiometer on the input to the LM386 for level control, we could create the finished schematic for the Waveform Generator (**Figure 18.11**). The finished project was soldered up on an Arduino protoshield and mounted in a SparkFun clear pcDuino/Arduino enclosure. The external switches, FREQUENCY SELECT potentiometer and output jack were mounted to the enclosure shell (**Figure 18.12**).

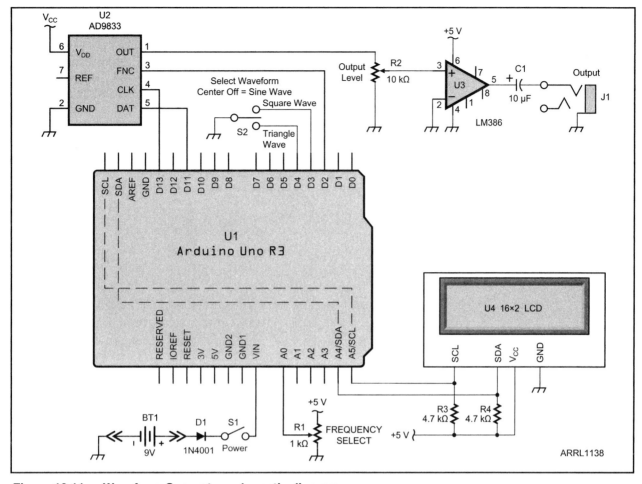

Figure 18.11 — Waveform Generator schematic diagram.

BT1 — 9 V battery.
C1 — 10 µF, 35 V electrolytic capacitor.
D1 — 1N4001 diode.
J1 — Stereo mini jack.
R1 — 1 kΩ potentiometer.
R2 — 10 kΩ 10-turn potentiometer.
R3, R4 — 4.7 kΩ, ⅛ W resistor.

S1 — SPST switch.
S2 — SPDT center-off switch.
U1 — Arduino Uno.
U2 — AD9833 programmable waveform generator breakout board.
U3 — LM386 audio power amplifier.
U4 — 16×2 I²C LCD.

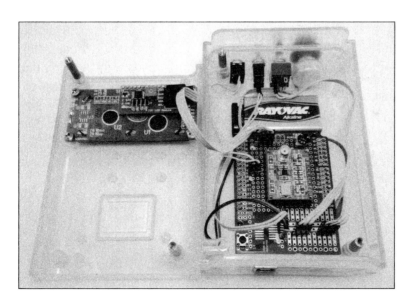

Figure 18.12 — Inside view of the finished Waveform Generator.

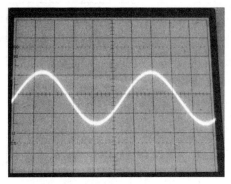

Figure 18.13 — Sine wave output of the AD9833 at 10 kHz. The waveform stayed clean all the way up to 12.5 MHz.

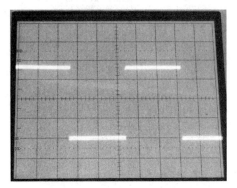

Figure 18.14 — Square wave output of the AD9833 at 10 kHz.

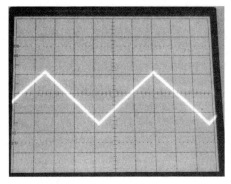

Figure 18.15 — Triangle wave output of the AD9833 at 10 kHz.

Once everything was mounted in the enclosure, some final tests were run to see how everything performed. By using a DDS module to generate our waveforms, the output waveforms were very crisp and distortion free across the entire 12.5 MHz operating range of the AD9833 DDS module (measured before the LM386 amplifier). Due to the coupling capacitor on the LM386, the edges of the square wave output started to lose their sharpness around 50 kHz, well above the 20 kHz point we set when the circuit was designed. **Figures 18.13** to **18.15** show sample waveforms.

The real fun came when I hooked up a frequency counter to see how accurate the DDS really was. They weren't lying about the 0.1 Hz resolution. The frequency output is dead on across the entire range as shown in the sample in **Figure 18.16**. So, what I thought was going to be a simple waveform generator has been transformed into a precision signal generator. You could actually use this circuit as the basis for a highly accurate VFO with a digital frequency display.

Enhancement Ideas

Since this project ended up way beyond being just an audio frequency waveform generator, there's a lot you can do with it. The potentiometer used to control the waveform generator has set limits, and if you increased the range of the potentiometer, you would lose your ability to accurately set a frequency. Replacing the potentiometer with a rotary encoder and an SPDT center-off switch would allow you to use the switch to select three different frequency stepping sizes (for example 1 Hz, 100 Hz and 10 kHz), and you could turn the rotary encoder continuously until you dialed in the exact frequency without having to turn it forever across the entire 12.5 MHz range. You could also add a control to adjust the phase shift of the output waveform. You could add a five position band switch and use this as the basis for an HF band VFO, or even a QRP transmitter. This is one of those projects that is only limited by your imagination. I started out being somewhat leery of the DDS modules, and now that I know how easy they really are to use, I can think of a whole list of projects I want to use with them.

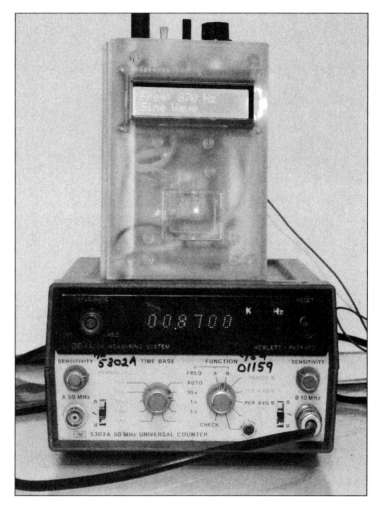

Figure 18.16 — Measuring the accuracy of the Waveform Generator.

References

Analog Devices — **www.analog.com**
Arduino Playground — **playground.arduino.cc**
RadioShack — **www.radioshack.com**
SparkFun Electronics — **www.sparkfun.com**
Texas Instruments — **www.ti.com**

CHAPTER 19

PS/2 CW Keyboard

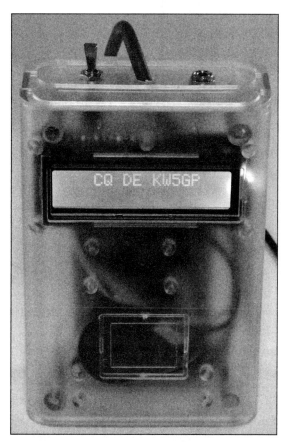

The finished PS/2 CW Keyboard mounted in a SparkFun pcDuino/Arduino enclosure.

As I've said earlier, CW has always been one of my first loves. However, over the years I've gotten away from it in favor of the digital modes, unless it's ARRL Field Day, then I'm planted somewhere on 20 or 40 meter SSB. But, I've recently wanted to get back into CW, and while my CW receiving speed is getting there, my sending "fist" leaves much to be desired. I don't get the usual "QLF" (Try sending with your left foot) reports. If there was such a thing, I'd more likely get something more like "QSF" (Are you sending with someone else's foot?) signal reports. I mean, my CW fist is bad, really, really bad.

Still, I enjoy working CW when I can, so I came up with the PS/2 CW Keyboard. This was actually one of the very first Arduino projects that I built. Originally, the project was built with an Ardweeny, but, as all early projects go, I wanted to upgrade it to an Arduino Uno in version 2 to make it look nicer, and to take advantage of the built-in voltage regulator on the Uno. Also, this was my

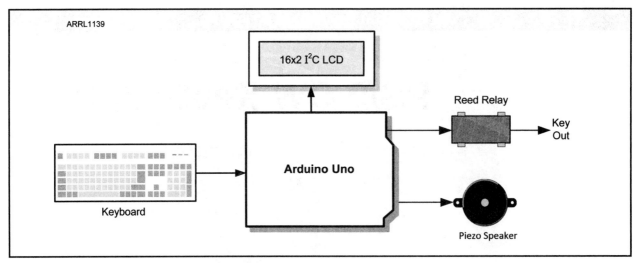

Figure 19.1 — PS/2 CW Keyboard block diagram.

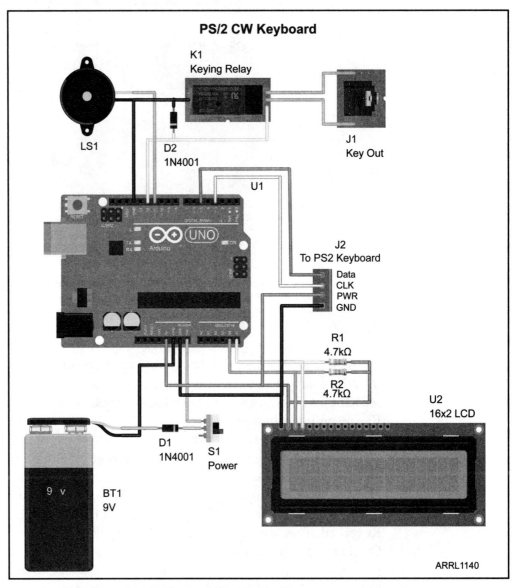

Figure 19.2 — PS/2 CW Keyboard Fritzing diagram.

chance to clean up the circuit board, as the Ardweeny version had been built on a standard piece of perfboard and was shoehorned into the project enclosure. Now, the new version of the project fits on an Arduino protoshield and looks much nicer.

For the PS/2 CW Keyboard, I wanted to use a standard PS/2-style keyboard, with variable sending speeds, and the ability to send either an audio tone to a speaker or plug into the standard key jack of a transmitter. I also wanted to be able to experiment with the EEPROM onboard the Arduino, so the PS/2 CW Keyboard would have five user-programmable 40 character memories that retain their information, along with the last selected sending speed and keying mode, after a reset or power-off. Finally, I also wanted to be able to turn the LCD backlight on or off via the keyboard. **Figure 19.1** shows the block diagram.

Now that we know what we want our PS/2 CW Keyboard to do, the next step is to assemble a circuit prototype on the breadboard as shown in **Figure 19.2**. The older PS/2-style keyboard was chosen for its simple interface and availability. For the PS/2-style keyboards, all you need to do is connect the Data and Clock lines, along with power and ground, from the PS/2 keyboard to the Arduino. The communication between the Arduino and the keyboard is handled by the `PS2Keyboard.h` library found in the Arduino Playground. For use with the PS/2 CW Keyboard sketch, the `PS2Keyboard.h` library had to be modified to support the F1-F12 keys, so when you compile your sketch, be sure to use the library in **Appendix A**, or download the modified library from **www.w5obm.us/Arduino**.

Creating the Sketch

We'll be introducing a couple of new features with this project. Since most modern CW keyers have memories, I wanted the PS/2 CW Keyboard to have memories as well. To accomplish this, we'll be using the built-in Arduino `eeprom.h` library to save our five memory function keys in the Arduino's onboard EEPROM memory, so that they will not be erased by a reset or power-off. As mentioned above, we'll also be using the `PS2Keyboard.h` library to interface with the PS/2-style keyboard.

Because we are using the Arduino's onboard 1K of EEPROM memory and building five 40 character CW memories, the logic for the PS/2 CW Keyboard sketch is a little more complex than the previous projects. Again, we'll start out with a flowchart (**Figure 19.3**) to break this down into manageable chunks. Fortunately, the libraries do most of the heavy lifting, so things are not really as difficult as they look at first glance.

The PS/2 CW Keyboard has multiple functions all controlled by keyboard commands. The left and right arrow keys are used to turn the LCD backlight on and off, while the up and down arrows are used to increase or decrease the CW sending speed by 1 word per minute (WPM). The F11 and F12 keys are used to select whether the CW is sent as an audio tone via a speaker or as a keying signal via the keying relay.

We have assigned the F1 through F5 keys as our CW memory (macro) keys. If we have previously saved data to these keys, the data will be sent each time the selected memory key is pressed. Each memory key can contain up to the

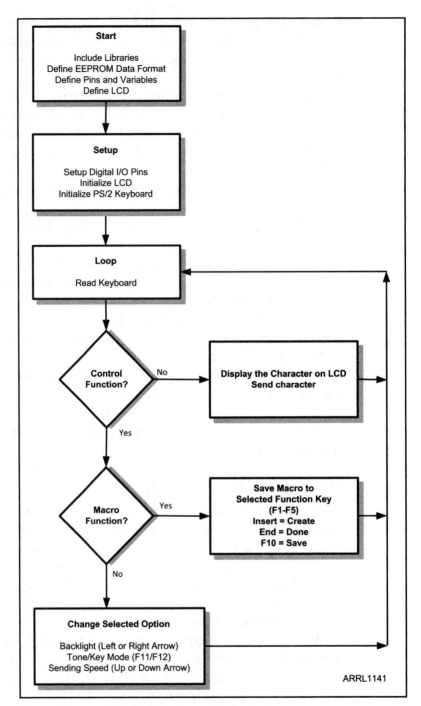

Figure 19.3 — PS/2 CW Keyboard flowchart.

number of characters defined by the bufflen variable minus 1 character. In this sketch, bufflen is set to 41. This means that you can have 40 characters in each of the five CW memories, with the last character in the entry reserved to indicate the end of the memory data. To create a memory key entry, press the INSERT key. This will display "Select Macro Key" on the LCD. Press the desired memory key (F1 through F5). The LCD will then display "Fx Macro Key Entry", where x will be the number of the function key you selected. Then, type the message you want to store to that CW memory key. When you have finished entering the message, press the END key, and the message will be assigned to the selected function key and saved in the Arduino's onboard EEPROM. When you save an entry to any memory function key, the current sending speed and keying mode are also saved to the EEPROM. Since the CW memory data, the sending speed and keying mode are stored in the Arduino's onboard EEPROM, they will be reloaded automatically any time you reset or power cycle your Arduino.

Also, if you want to save the current sending speed and keying mode at any time, just press the F10 key. This will store all of the current CW memories and operating parameters to the Arduino's onboard EEPROM.

Since the Arduino's EEPROM isn't part of the storage normally used for variable data, we have to manually handle the reading and writing of information to the EEPROM. The eeprom.h library does a lot of the work for us, but we still have to manually define the format needed for the data we'll be storing in the EEPROM. We'll also be defining our own command, the MIN statement, used to determine the minimum of two values.

```
#include <avr/eeprom.h>   // Include the AVR EEPROM Library

// Define the EEPROM data format
#define eeprom_read_to(dst_p, eeprom_field, dst_size) eeprom_read_block(dst_p,
(void *)offsetof(__eeprom_data, eeprom_field), MIN(dst_size, sizeof((__eeprom_
data*)0)->eeprom_field))
#define eeprom_read(dst, eeprom_field) eeprom_read_to(&dst, eeprom_field,
sizeof(dst))
#define eeprom_write_from(src_p, eeprom_field, src_size) eeprom_write_block(src_p,
(void *)offsetof(__eeprom_data, eeprom_field), MIN(src_size, sizeof((__eeprom_
data*)0)->eeprom_field))
#define eeprom_write(src, eeprom_field) { typeof(src) x = src; eeprom_write_
from(&x, eeprom_field, sizeof(x)); }

#define MIN(x,y) ( x > y ? y : x )   // Define a MIN function

#include <Morse.h>    // Include the Morse Library
#include <PS2Keyboard.h>   // Include the PS2Keyboard Library
#include <Wire.h>    // Include the Wire Communication Library
#include <LiquidCrystal_I2C.h>   // Include the LiquidCrystal I2C Library
```

The five CW memories used in this project each can hold up to 40 CW characters. When you add in all the other variables and constants used in this sketch, you'll find that we are right on the edge of running out of RAM for our variables. In fact, several times during the development of the sketch, strange things would start happening, and it turned out I had run out of variable memory space. Remember, the Arduino doesn't give you a convenient "Out Of Memory" error, so when you do run out of memory, your sketch will just start doing strange things. To help keep track of the memory usage, I added the get_free_memory() function from the Arduino Playground so that I could display the amount of free RAM still available. To use the get_free_memory(), we need to define some external variables in C.

```
extern int __bss_end;    // Used by Free Memory function
extern int *__brkval;
```

Next we'll define all the constants and variables used in the sketch:

```
const int buflen = 41; // Set size of Macros to 40 characters
const int lcd_end = 16; // set width of LCD
const int lcd_home = 0; // set the LCD home position
const int comm_speed = 9600; // Set the Serial port speed
const int current_id = 18; // EEPROM ID - used to verify EEPROM data is valid
const int lcd_address = 0x27; // I2C LCD Address
const int lcd_lines = 1; // Number of lines on LCD
const int beep_pin = 11; // Pin for CW tone
const int key_pin = 12; // Pin for CW Key
const int beep_on = 1; // 0 = Key, 1 = Beep

// Define the text data
const String EEPROM_invalid = "EEPROM data not valid";
const String ready = "Keyer Ready";
const String free_mem = "Free Mem:";
const String spd = "Speed = ";
const String f1key = "F1";
const String f2key = "F2";
const String f3key = "F3";
const String f4key = "F4";
const String f5key = "F5";
const String macro = " Macro";
const String selected = " Entry";
const boolean one = 1;
const boolean zero = 0;
const int start_speed = 15; // Starting CW speed
const String mode = " Mode";

// Set the PS2 keyboard I/O pins
const int DataPin = 5; // Set PS2 Keyboard Data Pin
const int IRQpin =  3; // Set PS2 Keyboard Clock Pin

char c;  // variable to hold the CW character to send
int key_speed = start_speed;  // the current keying speed
int cursor_pos = lcd_home;  // the current cursor position
int cursor_line = lcd_home;  // the current cursor line
boolean create_macro = zero;  // create macro flag
boolean macro_select = zero;  // the current macro being created
String macro_data;  // the current macro data
String add_data;  // the new macro data to replace
char macro_key;  // The macro entry key
String F1_data = "";  // The macro key data
String F2_data = "";
```

```
String F3_data = "";
String F4_data = "";
String F5_data = "";

// Boolean variable  to select keying or beep mode. 0 = key, 1 = beep
boolean morse_beep = one;

char xF[buflen];   // Temporary macro key data
String fdata;
int id;   // Our EEPROM id number read from the EEPROM
int morse_speed;   // The CW sending speed
```

Next, we'll define the structure of the data we'll be storing in EEPROM:

```
/*
 * __eeprom_data is the magic name that maps all of the data we are
 * storing in our EEPROM
 */
struct __eeprom_data      // The structure of the EEPROM data
{
  char eF1[buflen];   // the F1 macro
  char eF2[buflen];   // the F2 macro
  char eF3[buflen];   // the F3 macro
  char eF4[buflen];   // the F4 macro
  char eF5[buflen];   // the F5 macro
  int e_speed;   // the CW speed
  boolean e_beep;   // beep or key mode
  int EEPROM_ID;   // the EEPROM ID
  char * F;
};
```

Finally, we'll create the PS2Keyboard object as a 16-character by 1-line display. As you may recall from the Code Practice Oscillator project, the libraries for the 16-character by 2-line Hitachi HD44780-compatible displays don't handle two-line scrolling very well, so we'll be scrolling our data on just one line of the LCD. At this point, we'll also define a new Morse library object and set it to beep on pin 11, set the keying speed to the default of 15 WPM, and enable the speaker output.

```
PS2Keyboard keyboard;    // define the PS2Keyboard

// set the LCD address to 0x27 for a 16 chars and 1 line display
// The display is a 2 line display, but we only want scrolling on one line
LiquidCrystal_I2C lcd(lcd_address,lcd_end,lcd_lines);

//default to beep on pin 11, key speed to 15wpm and enable the speaker output
Morse morse(beep_pin, key_speed, beep_on);
```

In the `setup()` loop, we'll initialize the PS/2 keyboard and the LCD, and define the five memory buffers corresponding to the F1 through F5 keys on the PS/2 keyboard:

```
keyboard.begin(DataPin, IRQpin);   // Start the Keyboard
lcd.init(); // initialize the lcd
lcd.backlight();   // Turn on the LCD backlight
lcd.home();   // Go to Home position on LCD
lcd.print(ready);   // Show Ready on LCD
lcd.setCursor(lcd_end,lcd_home);   // Move to the end of Line 0
lcd.autoscroll();   // Enable Autoscroll
// Define the function keys
char F1[buflen], F2[buflen], F3[buflen], F4[buflen], F5[buflen];
```

In the last portion of the `setup()` loop, we'll read the contents of the EEPROM and save them into RAM:

```
eeprom_read(id,EEPROM_ID); // Read the EEPROM
// We put a set value in the EEPROM data to indicate the data is valid. If
// the set value is not there, the data is considered to be invalid
if (id != current_id)
{
  // Data not valid - keep the default settings
} else {
  // valid EEPROM DATA - Read EEPROM data
  eeprom_read(F1,eF1);   // Read the EEPROM data for the function keys
  eeprom_read(F2,eF2);
  eeprom_read(F3,eF3);
  eeprom_read(F4,eF4);
  eeprom_read(F5,eF5);

  // Read the EEPROM data for the current speed
  eeprom_read(morse_speed,e_speed);

// Read the EEPROM data for the current mode (Beep or Key)
  eeprom_read(morse_beep,e_beep);

  // Save EEPROM data to variables
  F1_data = String(F1);
  F2_data = String(F2);
  F3_data = String(F3);
  F4_data = String(F4);
  F5_data = String(F5);

  key_speed = morse_speed;   // Set the Key Speed
  mode_set();      // Set the Mode (beep or key)
}
```

In the main loop, we wait for a key to be pressed on the PS/2 keyboard, then use a `switch...case()` statement to determine what to do with the incoming characters. The first thing we do is check to see if it is a command that needs special handling. You'll notice that the logic is already included to decode most of the keyboard characters, but only a few are actually used.

```
// Check the keyboard to see if a key has been pressed
if (keyboard.available())
{
  c = keyboard.read();      // read the key

  // check for some of the special keys

  switch (c)  // Case Statement to determine which key was pressed
  {
      case PS2_ENTER:   // The ENTER key - Not Used
        break;

      case PS2_TAB:   // The TAB key - Not Used
        break;

      case PS2_ESC:   // The ESC key - Not Used
        break;

      case PS2_PAGEDOWN:   // The PageDown key - Not Used
        break;

      case PS2_PAGEUP:   // The PageUp key - Not Used
        break;

      case PS2_LEFTARROW:   // The Left Arrow key
        lcd.noBacklight(); // Turn the LCD backlight off
        break;
      case PS2_RIGHTARROW:   // The Right Arrow key
        lcd.backlight(); // Turn the LCD backlight on
        break;

      case PS2_UPARROW:   // The Up arrow key
        key_speed = key_speed ++;  // Increase CW Speed by 1wpm
        lcd.noAutoscroll();  // Clear the LCD and display current speed
        lcd.clear();
        lcd.home();
        lcd.print(spd + String(key_speed));
        lcd.setCursor(lcd_end,lcd_home);
        lcd.autoscroll();
        increase_speed();  // increase the CW speed by 1wpm
        break;
```

```
      case PS2_DOWNARROW:   // The Down Arrow key
         key_speed = key_speed --;    // decrease CW speed by 1wpm
         lcd.noAutoscroll();// Clear the LCD and display current speed
         lcd.clear();
         lcd.home();
         lcd.print(spd + String(key_speed));
         lcd.setCursor(lcd_end,lcd_home);
         lcd.autoscroll();
         decrease_speed();   // decrease the CW speed by 1wpm
         break;

      case PS2_DELETE:   // The DELETE key
         morse.sendmsg("EEE");// Send 3 dits for error
         break;
```

We handle the F1 through F5 keys differently because they have differing functions depending on whether you are sending the CW memory data assigned to that key, or are in the process of assigning CW memory data, referred to in the sketch as the macro data, to the selected function key.

```
case PS2_F1:   // Macro Key F1
  // If the F1 key has a macro assigned to it and we are not
  // currently creating a macro, send the macro via CW
  if (F1_data !="" and !create_macro)
  {
    send_macro(F1_data);   // Send the Macro data if we're not creating it
    break;
  }
  if (create_macro)   // If we're creating the macro for F1
  {
    // Check Macro_select and select this key if not already selected
    if (macro_select)
    {
      // We've already selected it, break out - we want to add data
      break;
    } else {
      // Select F1 Macro
      lcd.noAutoscroll();
      lcd.clear();
      lcd.home();
      lcd.print(f1key + macro + selected);   // Display the macro key selected
      lcd.setCursor(lcd_end,lcd_home);
      lcd.autoscroll();
      macro_select = one;   // A macro key has been selected
      macro_key = PS2_F1;   // The macro key we're creating is F1
    }
    else {
    break;
  }
  break;
```

We then duplicate this operation for the F2 thru F5 keys:

```
case PS2_F2:  // Macro Key F2
if (F2_data !="" and !create_macro)
{
  send_macro(F2_data);  // Send the Macro data if we're not creating it
  break;
}
if (create_macro)
{
  // Check Macro_select and select this key if not already selected
  if (macro_select)
  {
    // We've already selected it, break out - we want to add data
   break;
  } else {
    // Select F2 Macro
    lcd.noAutoscroll();
    lcd.clear();
    lcd.home();
    lcd.print(f2key + macro + selected);  // Display the macro key selected
    lcd.setCursor(lcd_end,lcd_home);
    lcd.autoscroll();
    macro_select = one;  // A macro key has been selected
    macro_key = PS2_F2;  // The macro key we're creating is F2
  }
} else {
  break;
}
break;

case PS2_F3: // Macro Key F3
  if (F3_data !="" and !create_macro)
  {
    send_macro(F3_data);// Send the Macro data if we're not creating it
    break;
  }
  if (create_macro)
  {
    // Check Macro_select and select this key if not already selected
    if (macro_select)
    {
      // We've already selected it, break out - we want to add data
      break;
    } else {
      // Select F3 Macro
      lcd.noAutoscroll();
      lcd.clear();
```

```
          lcd.home();
          lcd.print(f3key + macro + selected);   // Display the macro key selected
          lcd.setCursor(lcd_end,lcd_home);
          lcd.autoscroll();
          macro_select = one;   // A macro key has been selected
          macro_key = PS2_F3;   // The macro key we're creating is F3
        }
    } else {
      break;
    }
    break;

case PS2_F4:
  if (F4_data !="" and !create_macro)
  {
    send_macro(F4_data);   // Send the Macro data if we're not creating it
    break;
  }
  if (create_macro)
  {
  // Check Macro_select and select this key if not already selected
  if (macro_select)
  {
    // We've already selected it, break out - we want to add data
    break;
  } else {
    // Select F4 Macro
    lcd.noAutoscroll();
    lcd.clear();
    lcd.home();
    lcd.print(f4key + macro + selected);   // Display the macro key selected
    lcd.setCursor(lcd_end,lcd_home);
    lcd.autoscroll();
    macro_select = one;   // A macro key has been selected
    macro_key = PS2_F4;   // The macro key we're creating is F4
  }
} else {
  break;
}
break;

case PS2_F5:
  if (F5_data !="" and !create_macro)
  {
    send_macro(F5_data);   // Send the Macro data if we're not creating it
    break;
  }
  if (create_macro)
  {
```

```
    // Check Macro_select and select this key if not already selected
    if (macro_select)
    {
      // We've already selected it, break out - we want to add data
      break;
    } else {
      // Select F5 Macro
      lcd.noAutoscroll();
      lcd.clear();
      lcd.home();
      lcd.print(f5key + macro + selected);  // Display the macro key selected
      lcd.setCursor(lcd_end,lcd_home);
      lcd.autoscroll();
      macro_select = one;   // A macro key has been selected
      macro_key = PS2_F5;   // The macro key we're creating is F5
    }
  } else {
    break;
  }
  break;
```

Again, the F6 through F9 keys are decoded, but do not have any actions assigned to them:

```
case PS2_F6:   // The F6 key - Not Used
  break;

case PS2_F7:   // The F7 key - Not Used
  break;

case PS2_F8:   // the F8 key - Not Used
  break;

case PS2_F9:   // The F9 key - Not Used
  break;
```

We can use the F10 key to save our current CW memory data, sending speed and keying mode to the EEPROM at any time:

```
case PS2_F10:   // The F10 key
  save_macro();   // Save all data to the EEPROM
  break;
```

We use the F11 and F12 keys to switch between the beep and keying modes:

```
case PS2_F11: // F11 key - Turn on Beep Mode
  lcd.noAutoscroll();
  lcd.clear();
  lcd.home();
```

```
    lcd.print("Beep" + mode);   // Clear the LCD and display Beep Mode
    lcd.setCursor(lcd_end,lcd_home);
    lcd.autoscroll();
    morse_beep = one;
    mode_set();   // Set the Mode to Beep
    break;

case PS2_F12: // F12 key - Turn on Key Mode
    lcd.noAutoscroll();
    lcd.clear();
    lcd.home();
    lcd.print("Key" + mode);   // Clear the LCD and display Keying Mode
    lcd.setCursor(lcd_end,lcd_home);
    lcd.autoscroll();
    morse_beep = zero;
    mode_set();   // Set the Mode to Keying
    break;
```

The INSERT key is used to initiate the process of creating a CW memory entry:

```
case PS2_INSERT: // INSERT key - Enter Macro Key Data
    // If we want to create a macro but haven't selected a key yet
    if (!create_macro and !macro_select)
    {
      lcd.noAutoscroll();
      lcd.clear();
      lcd.home();
      lcd.print("Select" + macro + " Key");   // Display F-Key Selection Message
      lcd.setCursor(lcd_end,lcd_home);
      lcd.autoscroll();
      create_macro = one;   // Set the Create macro flag
      macro_data ="";   // Clear the macro data
    }
    break;
```

The END key is used to signify the end of a macro entry:

```
case PS2_END: // END key - End Macro Data Entry
    // If we're creating a macro entry, check to see if a key was selected
    if (create_macro)
    {
      // Abort or Save the Macro
      lcd.noAutoscroll();
      lcd.clear();
      lcd.home();
      if (macro_select)   // If a macro key has been selected, save it
      {
```

```
      save_macro();  // Save the Macro data to EEPROM
      lcd.print(macro + " Saved");  // Display "Saved" on the LCD
    } else {
      // Aborting - Don't save and display "Aborted" on the lCD
      lcd.print(macro + " Aborted");
    }
  }
  // We're all done with the Macro, clear the screen and reset the flags
  lcd.setCursor(lcd_end,lcd_home);
  lcd.autoscroll();
  create_macro = zero;
  macro_select = zero;
  break;
```

As the last step in the `switch...case()` statement, the default action is to do nothing and continue on through the loop:

```
default:   // Otherwise do the default action, which is to continue on
```

As the last step in the main `loop()`, we handle the building of the CW memory entry if we are creating one, check to see if we've ended the creation of a CW memory entry, or just send the character normally.

```
if (create_macro) // If we're creating a macro
{
  // check to see if we have selected key
  if (macro_select) // If we have a valid key, get the macro data
  {
    lcd.noAutoscroll();
    lcd.clear();
    lcd.home();
    // Display that we're ready to enter Macro data for the selected key
    lcd.print("Enter" + macro);
    c = toupper(c);  // convert the character to uppercase

    // As long as the macro is <= max length, add to the macro data
    if (macro_data.length()<(buflen - 1))
    {
      macro_data = String(macro_data) + String(c);
    }
    lcd.setCursor(lcd_end,lcd_home);
    lcd.autoscroll();
  } else {
    // Select the Macro Key
    if (c == PS2_END)   // Check to see if we want out
    {
      macro_select = zero; // Reset the Macro Flags
      create_macro = zero;
    }
```

```
     }
   } else {
     // If we're not doing the Macro thing just print the character to the LCD
     // and send all normal characters

     lcd.write(toupper(c)); // Display the character on the LCD

     morse.send(c);   // Send the character
   }
```

In the main `loop()`, we call one of two functions to increase or decrease the CW sending speed. As you can see, the `Morse.h` library actually handles all the speed changes for us.

```
int increase_speed()   // Function to Increase CW Speed by 1wpm
{
  if (morse_beep)
  {
    Morse morse(beep_pin, key_speed, one); //beep
  } else {
    Morse morse(key_pin, key_speed, zero); // no beep
  }
  return key_speed;
}

int decrease_speed()    // Function to decrease CW Speed by 1wpm
{
  if (morse_beep) {
    Morse morse(beep_pin, key_speed, one); // beep
  } else {
    Morse morse(key_pin, key_speed, zero); // keyer
  }
  return key_speed;
}
```

The `save_macro()` function is used to save the CW memory information to the specified memory function (macro) key and to the Arduino's onboard EEPROM.

```
// Function to Save the macro to variable - and write it to EEPROM
void save_macro()
{
  // Trim the function key data if longer than the memory key buffer
  if (macro_data.length() > (buflen - 1))
  {
    macro_data = macro_data.substring(0,(buflen - 1));
  }
  switch (macro_key) // Determine which key to save the data to
```

```c
{
  case PS2_F1:
    F1_data = macro_data;
    break;

  case PS2_F2:
    F2_data = macro_data;
    break;

  case PS2_F3:
    F3_data = macro_data;
    break;

  case PS2_F4:
    F4_data = macro_data;
    break;

  case PS2_F5:
    F5_data = macro_data;
    break;
}

// The EEPROM ID byte is used to signify the EEPROM has valid dats
id = 18;  // set the EEPROM ID Byte
morse_speed = key_speed; // set the keying speed
morse_beep = 1; // set the beep mode
fdata = F1_data;
m_data();  // Format the data for writing to EEPROM
eeprom_write_from(xF, eF1,buflen);  // Write the Data to EEPROM

fdata = F2_data;
m_data();  // Format the data for writing to EEPROM
eeprom_write_from(xF, eF2,buflen);  // Write the Data to EEPROM

fdata = F3_data;
m_data();  // Format the data for writing to EEPROM
eeprom_write_from(xF, eF3,buflen);  // Write the Data to EEPROM

fdata = F4_data;
m_data();  // Format the data for writing to EEPROM
eeprom_write_from(xF, eF4,buflen);  // Write the Data to EEPROM

fdata = F5_data;
m_data();  // Format the data for writing to EEPROM
eeprom_write_from(xF, eF5,buflen);  // Write the Data to EEPROM
```

```
  eeprom_write(morse_speed,e_speed);   // Write the current speed to EEPROM
  eeprom_write(morse_beep,e_beep);     // Write the current mode to EEPROM

  eeprom_write(id,EEPROM_ID);   // Write the EEPROM ID to EEPROM
  fdata = "";
}
```

The m_data() function is used to convert our memory key variable data to a format that can be written to and read from the EEPROM.

```
void m_data()    // Function to convert data to EEPROM writable format
{
  // This routine figures out where to write the variable data in the EEPROM
  char yF1[buflen];
  int i = 0;
  if (fdata.length() > (buflen - 1)) // trim the data to the right size
  {
    fdata = fdata.substring(0,(buflen - 1));
  }
  // if the variable has data, assign it to the variable one
  // character at a time for each Macro key
  if (fdata.length() > 0)
  {
    while (i <= fdata.length() -1)
    {
      yF1[i] =  fdata.charAt(i);
      i++;
    }
    yF1[i] = 0;
    i = 0;
    // place the data in the variable one character at time
    // to write to the EEPROM
    while (yF1[i] != 0)
    {
      xF[i] = yF1[i];
      i++;
    }
  }
  xF[i] = 0;
}
```

The send_macro() function is used to send the CW memory (macro) data when any of the CW memory keys (F1 through F5) are pressed.

```
void send_macro(String F_Key_data)    // Function to Send the Macro data
{
  char send_data[buflen];
  int i = 0;

  lcd.print(F_Key_data); // Display the Memory Key data on the LCD
  // Fill the character array with each character in the macro
  // one character at a time
  while (i <= F_Key_data.length() -1)
  {
    send_data[i] = F_Key_data.charAt(i);
    i++;
  }
  // Add a zero at the end of the array to signify end of text
  send_data[i] = 0;
  // send the data in the send_data array as a message
  morse.sendmsg(send_data);
}
```

The `mode_set()` function is used to select either audio tone or keying mode:

```
void mode_set()    // Function to Set the mode to beep or keying
{
  if (morse_beep)
  {
    Morse morse(beep_pin, key_speed, one); //default to beep on pin 11
  } else {
    Morse morse(key_pin, key_speed, zero); //default to key on pin 12
  }
}
```

Finally, as a debugging aid, we include the `get_free_memory()` function from the Arduino Playground to give us the amount of free RAM remaining:

```
int get_free_memory()   // Function to get free memory
{
  int free_memory;

  if((int)__brkval == 0)
    free_memory = ((int)&free_memory) - ((int)&__bss_end);
  else
    free_memory = ((int)&free_memory) - ((int)__brkval);

  return free_memory;
}
```

This project was my first experience with the `Morse.h` and

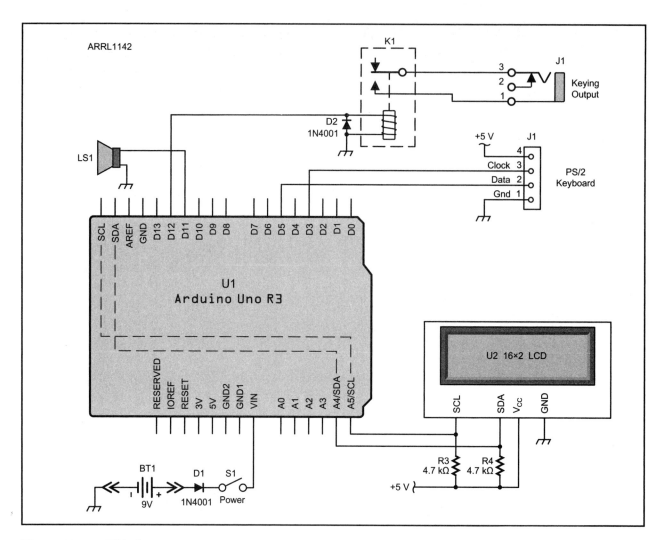

Figure 19.4 — PS/2 CW Keyboard schematic diagram

BT1 — 9 V battery.
D1, D2 — 1N4001 diode.
J1 — ⅛ inch stereo mini jack.
K1 — 5 V SPST reed relay.
LS1 – 8 Ω mini speaker.

R1, R2 — 4.7 kΩ, ⅛ W resistor.
S1 — SPST switch.
U1 — Arduino Uno.
U2 — 16×2 serial I²C LCD.
PS/2 keyboard

PS2Keyboard.h libraries. They were also the first libraries I had to dig into and modify. The Morse.h library had a couple of encoding errors with some of the Morse code characters, which were easily resolved. The PS2Keyboard.h library did not support some of the extended characters such as INSERT, END, and the F1 through F12 Keys. It was relatively easy to

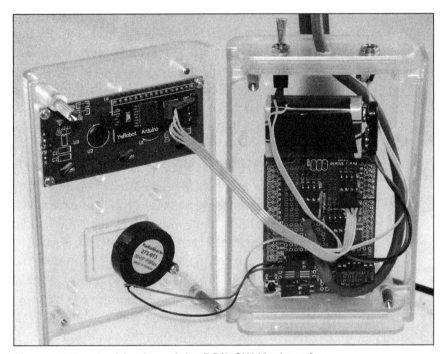

Figure 19.5 — Inside view of the PS/2 CW Keyboard.

Figure 19.6 — The flexible keyboard I use with the PS/2 CW Keyboard for portable operation.

Figure 19.7 — The PS/2 CW Keyboard unit rolled inside the flexible keyboard, ready for next Field Day.

make the necessary changes in the library to add the necessary functions. The moral of the story is, don't be afraid if a library doesn't do exactly what you need it to, or if it has a bug. Most libraries are almost as easy to troubleshoot as Arduino sketches and the majority of your standard Arduino troubleshooting techniques can be used while working with libraries.

Once I had everything working and debugged on the breadboard, the protoshield was wired up according the schematic diagram in **Figure 19.4**, and the finished project was mounted in a SparkFun pcDuino/Arduino project enclosure (**Figure 19.5**). I was able to find an older flexible keyboard (**Figure 19.6**) with a USB connector that would work with a USB to PS/2 adapter, allowing the whole project to be rolled up inside the keyboard (**Figure 19.7**) and thrown into the Field Day kit. The newer USB keyboards won't work with the USB to PS/2 adapter, so you'll need to dig up an older USB keyboard or add a USB host shield if you want to use a USB keyboard with this project.

Enhancement Ideas

With the low number of external components, this project would easily fit inside a smaller enclosure such as a mint tin or similar small enclosure. You could use an Arduino Nano and an organic LED (OLED) display to shrink this into a project you could carry in your pocket. If you'd prefer not to use a PS/2-style keyboard, you could use an Arduino Uno with the USB host shield and a standard USB keyboard. You could also optimize the memory utilization and give yourself more CW memory function keys by moving all of the constant variables into the Arduino flash memory using the `F()` macro or the `PROGMEM` statement with the `avr/pgmspace./h` library.

And the best part of this whole project: now I can send near-perfect CW and nobody has to know how bad my CW sending ability really is. I've had a lot of fun with the PS/2 CW Keyboard in contests, where I've had a lot of the required contest exchange and CQs stored in the memory function keys, allowing me time to log the contact on the computer while the keyboard is off sending CQ, looking for the next contact.

References

Arduino — **www.arduino.cc**
Arduino Playground — **playground.arduino.cc**
CadSoft — **www.cadsoftusa.com**
Fritzing — **www.fritzing.org**
SparkFun Electronics — **www.sparkfun.com**

CHAPTER 20

Field Day Satellite Tracker

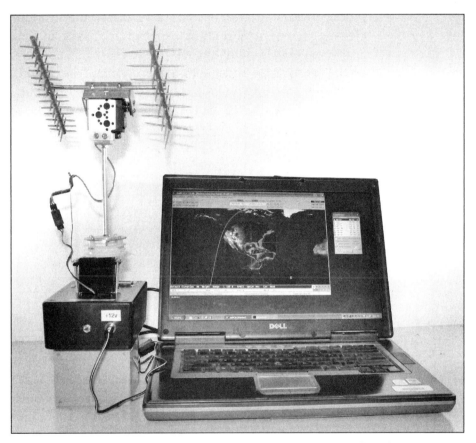

The finished Field Day Satellite Tracker being controlled by *SatPC32*.

The Field Day Satellite Tracker was the second major Arduino project I attempted. The Satellite Tracker described here provides azimuth and elevation rotation for a pair of model satellite antennas to demonstrate how the tracking software and automatic antenna positioning work. The Satellite Tracker uses *SatPC32* satellite tracking software and follows actual amateur satellites, just like a real antenna system. It was originally created as a proof-of-concept prototype for a full-scale satellite azimuth and elevation controller for my Yaesu G5400 satellite antenna rotator. Somewhere along the way, the project took on a life of its own and our club has used it at many of our outdoor and public events such as ARRL Field Day and ARRL Kids Day.

At our club's last ARRL Kid's Day event, we had both the Satellite Tracker and a full-size satellite antenna array up and running with its own Arduino-based controller, tracking satellites side by side. It was an awesome sight to see the little tracker start to move and look up to see the real antenna tracking satellites right along with it. Whenever the tracker started to move, everyone made their way to the satellite station's tent to hear us trying to work the satellites.

This is yet another project that shows the versatility of the Arduino. The Satellite Tracker interfaces with a PC running the *SatPC32* software to do the actual tracking, and uses the RS-232 satellite rotator control output from the tracking software to drive a pan/tilt servo assembly that moves the antenna model mounted on top of the pan/tilt servo head. As far as the PC software is concerned, it thinks it's communicating with the Yaesu GS-232A azimuth/elevation (az/el) rotator computer interface, when in actuality it is talking to an Ardweeny that is emulating the Yaesu GS-232A and driving the servos based on the rotator movement commands coming from the PC. We'll also display the current positioning information on a 16-character by 2-line LCD display mounted in the base of the enclosure. **Figure 20.1** shows the block diagram.

We'll be using a pair of servos similar to the one shown in **Figure 20.2** to

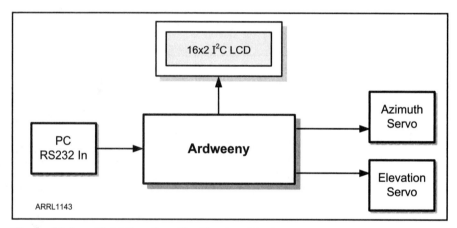

Figure 20.1 — Field Day Satellite Tracker block diagram.

Figure 20.2 — The servos used in the Satellite Tracker's pan/tilt servo assembly.

rotate and tilt our antenna model. Servos are small, powerful, geared motor assemblies often used in robotics and radio controlled models that use the width of a pulse on a digital I/O pin to determine how far they should rotate. Servos are very accurate and move quickly to the desired position. Your typical servo has a range of travel from 0 to 180°. This is fine for elevation rotator positioning, but we need a range of 0 to 360° for azimuth rotation if we're going to get the PC to think we're a real rotator. Since the standard servos don't have that range of motion, we'll use a model sailboat multiple turn "sail winch" servo. The sail winch servo works on the same principle as a regular servo, except that it can turn up to seven full rotations, which makes it perfect for the 0 to 360° rotation we'll need for the azimuth portion of the Satellite Tracker.

Circuit Highlights

This project is built using a Solarbotics Ardweeny, shown in **Figures 20.3 and 20.4**. The Ardweeny uses the same Atmel ATmega328 processor used in the Arduino Uno, except the supporting components are mounted on a backpack-style circuit board that is soldered directly to the pins on the ATmega. This gives you the equivalent of an Arduino Uno that has a very small footprint and fits in a standard 28 pin IC socket. The Ardweeny does not have a USB

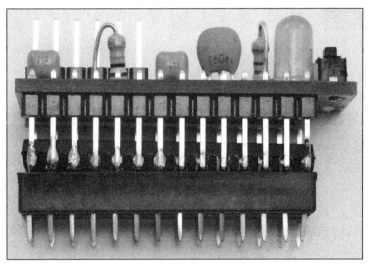

Figure 20.3 — The Solarbotics Ardweeny.

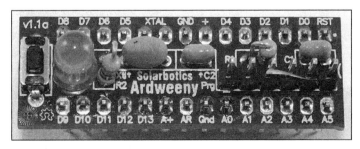

Figure 20.4 — Top view of the Ardweeny showing the pin layouts.

port for programming. Instead, the Ardweeny requires the FTDI USB-to-TTL serial module (**Figure 20.5**) to allow your workstation to communicate with the Arduino. Since the PC I would be using to run the *SatPC32* satellite tracking software had an RS-232 port, I chose to use the Ardweeny so that I could connect the Ardweeny to the PC's RS-232 port using a MAX232 RS-232-to-TTL converter chip. Lastly, the Ardweeny does not have an onboard power regulator, so we will also need an external source of 5 V to power the project.

Figure 20.6 shows the Fritzing diagram for the Field Day Satellite Tracker project. The PC interfaces to the Ardweeny using a MAX232 RS-232-to-TTL

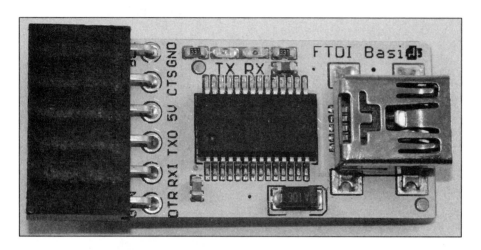

Figure 20.5 — The FTDI USB-to-serial adapter module.

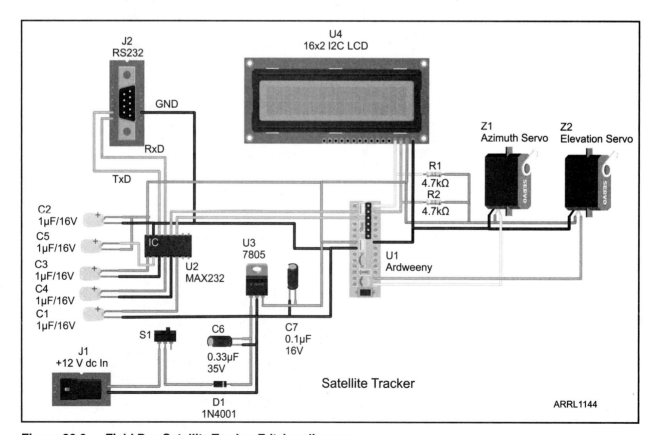

Figure 20.6 — Field Day Satellite Tracker Fritzing diagram.

serial converter that attaches to the hardware serial port on the Ardweeny. For power, a small 12 V, 5 Ah gel cell battery attaches to the bottom of the project enclosure using hook-and-loop fastener strips and plugs into a dc power jack on the side of the enclosure. This makes the entire project portable and self-powered, which is perfect if you use a laptop PC to run the tracking software. The servos are controlled by a 5 V pulse on the data line, so they can be driven directly from the Ardweeny's digital I/O pins. We'll finish things up using a 16-character by 2-line LCD display with an I²C "backpack" that communicates with the Ardweeny using the I²C bus.

The Sketch

With everything assembled on the breadboard, we can begin to write the sketch. When you break the sketch down into a flowchart (**Figure 20.7**), you'll see this really isn't a difficult sketch to create. Since the Satellite Tracker looks exactly like a Yaesu GS-232A azimuth/elevation rotator controller as far as the PC software is concerned, the key is determining the format of the RS-232 commands the PC will send to the Ardweeny. The Ardweeny will decode the desired azimuth and elevation information from the RS-232 commands and translate the commands into servo positioning information. This is done by mapping the desired azimuth and elevation from degrees into pulse widths to drive the servos. The Arduino IDE's built-in Servo.h library will handle the actual positioning of the servos based on the pulse width needed to rotate the

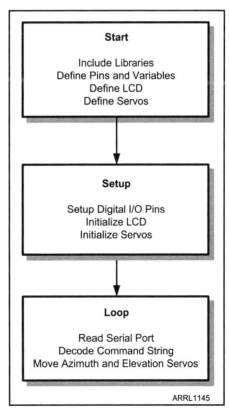

Figure 20.7 — Field Day Satellite Tracker flowchart.

servos to the desired position.

Starting out with the sketch, we'll include the `Servo.h`, `Wire.h` and `LiquidCrystal_I2C` libraries we'll need for the servos and the LCD display. We'll also initialize the LCD and servo objects:

```
#include <Servo.h>    // Use the Servo Library
#include <Wire.h>     // Use the I2C Communication Library
#include <LiquidCrystal_I2C.h>   // Use the LiquidCrystal_I2C Library

LiquidCrystal_I2C lcd(0x27,16,2);   // set the LCD address to 0x27 for a 16 chars
and 2 line display

Servo myservoAZ;    // create servo object to control Azimuth servo
Servo myservoEL;    // create servo object to control Elevation servo
```

Next we'll define the variables and constants we'll need for the sketch. The values used for the servo positions are determined manually and will be different from one servo to another. Adjust these values until your servos are positioning correctly. The *SatPC32* software has the ability to send positioning commands manually, so you can command the servos to go to 0 and 360° and modify the values accordingly.

```
// Set the Azimuth Servo pulse width for 0 Degrees
int min_servo_pulse = 900;

// Set the Azimuth Servo pulse width for 360 Degrees
int max_servo_pulse = 1970;

// Set the Elevation Servo pulse width for 0 Degrees
int EL_Min_pulse = 640;

// Set the Azimuth Servo pulse width for 180 Degrees
int EL_Max_pulse = 2340;

int delay1 = 50;
int delay2 = 5000;
int currentEL = 0;   // Variable to hold the current elevation
int currentAZ = 0;   // variable to hold the current azimuth
int inByte;   // the incoming byte on the Serial port

// Flag to indicate we are decoding a move command
boolean move_command = false;

int byte_count = 0;   // the number of bytes received on the Serial port
char az_buffer[4];    // Holds the decoded azimuth command
char el_buffer[4];    //  Holds the decoded elevation command
char rotor_buffer[10];   // The incoming Rotor command data
int set_az, set_el;   // Holds the desired azimuth and elevation data
```

In the `setup()` loop we initialize the LCD and have it display a brief startup message so we know everything is working. Then we'll initialize the servos, start the serial port for communication with the PC, and then move both servos to the 0° "home" position.

```
  lcd.init(); // initialize the LCD
  lcd.backlight();  // Turn on the LCD backlight
  lcd.setCursor(0,1);
  lcd.print("KW5GP SatTrack");
  lcd.setCursor(0, 0);
  lcd.print("Rotor Controller");
  delay(delay2);

// attaches the servo on pin 9 to the Azimuth servo object
  myservoAZ.attach(9);

// attaches the servo on pin 10 to the Elevation servo object
  myservoEL.attach(10);

  pinMode(13, OUTPUT);   // initialize the pin 13 as an output.

  Serial.begin(9600);   // Start the Serial Port at 9600 baud

  lcd.clear();
  lcd.setCursor(0, 0);
  lcd.print("Homing Servos    ");
  set_azimuth(0);   // Set the Servos to 0 degrees
  set_elevation(0);
  lcd.setCursor(0,1);
  lcd.print("Elevation:  0");  // Display the current Elevation
  lcd.setCursor(0, 0);
  lcd.print("Azimuth  :  0");  // Display the current Azimuth
  delay(delay2);
```

In the main `loop()`, we wait until a character is received on the serial port connected to the PC:

```
if (Serial.available())     // read data from the Serial Port
  {
    inByte = Serial.read();
    Serial.write(inByte);
```

We then check to see if the first character is a "W," which is the Yaesu GS-232A command to move the rotator to the desired azimuth and elevation. The full command we'll be decoding is "Waaa eee," where aaa is the desired azimuth in degrees, and eee is the desired elevation in degrees. If it is a "W," we enable the sketch logic to decode the rest of the command string.

Field Day Satellite Tracker

```
if (inByte == 87) // If the incoming data is a "W", it's a rotor move command
  {
    // Turn on string build - next 7 chars = azimuth and elevation
    move_command = true;
    byte_count = 0;
  }
```

If it is not a "W," we'll start building the strings for the azimuth and elevation data from the incoming characters if we have already received a "W." If not, the character is ignored and the sketch continues to monitor the serial port waiting for the "W" command. We continue to build the azimuth and elevation strings until we have received a total of eight characters, the length of the complete move command.

```
//  If it's not a move command - add to string if we're building
    if (move_command and inByte !=87 )
    {
      rotor_buffer[byte_count] = char(inByte);
      byte_count = byte_count + 1;
      if (byte_count <= 3)   // The first 3 characters are Azimuth
      {
        az_buffer[byte_count - 1] = char(inByte);
      }

      // The last 3 characters are Elevation
      if (byte_count >= 5 and byte_count <= 7)
      {
        el_buffer[byte_count -5] = char(inByte);
      }

      // Once we have 8 characters, we have a complete move command
      if (byte_count == 8) {
       // Convert the Azimuth data to an integer
        set_az = atoi(az_buffer);
       // Convert the Elevation data to an integer
        set_el = atoi(el_buffer);

        // Move the Azimuth Servo to desired position
        set_azimuth(set_az);
        lcd.setCursor(11,0);
        lcd.print(set_az);   // Display the New Azimuth on the LCD
        lcd.print("   ");

        // Move the Elevation Servo to desired position
        set_elevation(set_el);
        lcd.setCursor(11,1);
        lcd.print(set_el);   // Display the New Elevation
        lcd.print("   ");
```

```
        move_command = false;  // Indicate we're done with this move command
    }
}
```

The servo positioning is performed by the `set_azimuth` and `set_elevation` functions found after the main `loop()` in the sketch. These functions will use the `map()` statement to convert the decoded positioning information from degrees into the pulse widths the servos need to perform the actual positioning.

```
// Set Azimuth Funtion
// Moves the Azimuth Servo to desired position
void set_azimuth(int desired_az)
{
  int move_azimuth;

  // Map the desired position to the correct Azimuth servo pulse width
  move_azimuth = map(desired_az,360, 0,min_servo_pulse,max_servo_pulse);
  // Send the Servo position pulse to the Azimuth servo
  myservoAZ.writeMicroseconds(move_azimuth);
  delay(delay1);  // Wait for move to complete
}


// Set Elevation Function
// Moves the Elevation Servo to desired position
void set_elevation(int desired_el)
{
 int move_elevation;
 // Map the desired position to the correct Elevation servo pulse width
 move_elevation = map(desired_el,0, 180,EL_Min_pulse,EL_Max_pulse);

 // Send the Servo position pulse to the Elevation servo
 myservoEL.writeMicroseconds(move_elevation);
 delay(delay1);  // Wait for move to complete
}
```

Once everything was working on the breadboard, the schematic for the circuit (**Figure 20.8**) was drawn up, the circuit was soldered up on a perfboard, and the finished board was mounted inside a 2 × 4 × 6 inch RadioShack project enclosure as shown in **Figures 20.9** and **20.10**. The 16-character by 2-line LCD is mounted inside the top lid of the enclosure along with the sail winch servo for the azimuth rotation. The pan/tilt servo assembly for elevation is mounted to the sail winch servo using a 4 inch section of threaded aluminum tubing I found at a local hobby shop. A model of a satellite antenna array was built out of wooden dowels and toothpicks, painted silver to make it look like the real thing, and mounted on top of the pan/tilt servo assembly. **Figure 20.11** shows the completed Satellite Tracker with its model antennas.

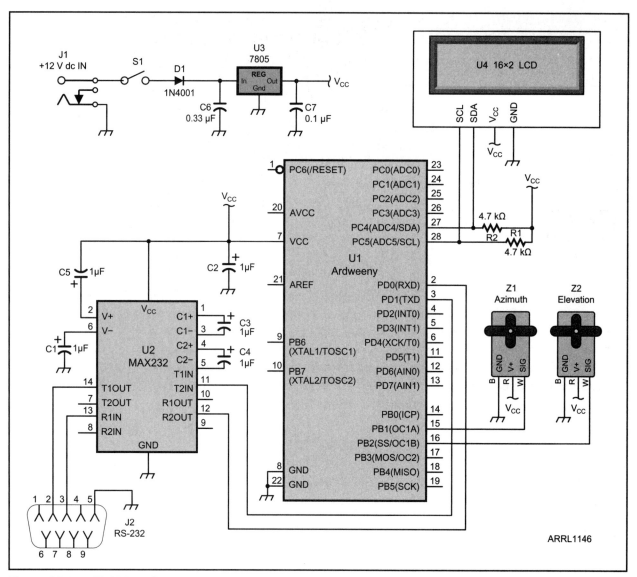

Figure 20.8 — Field Day Satellite Tracker schematic diagram.

C1-C5 — 1 µF, 16 V capacitor.
C6 — 0.33 µF, 35 V capacitor.
C7 — 0.1 µF, 16 V capacitor.
D1 — 1N4001 diode.
J2 — DB9 male connector.
J1 — DC power jack.
R1, R2 — 4.7 kΩ, ⅛ W resistor.
S1 — SPST toggle switch.

U1 — Solarbotics Ardweeny.
U2 — MAX232 RS-232 transceiver.
U3 — LM7805 5 V regulator.
U4 — 16×2 I²C LCD.
Z1 — 7 turn sail winch servo.
Z2 — Small standard servo.
Pan/tilt servo bracket
Radio Shack project enclosure (270-1806)

On the PC side of things, all you have to do is configure *SatPC32*'s rotator setup to use a Yaesu GS-232A on the desired serial port at 9600 baud, connect up the Field Day Satellite Tracker, and you should be good to go. Be sure to select the Yaesu GS-232A and not the GS-232B as the GS-232B uses a slightly different format for the positioning commands.

If you get a serial port error and *SatPC32* is trying to connect to the wrong serial port, there's no easy way to change the serial port in the software. You'll need to edit the Documents and Settings\%User%\Application Data\SatPC32\

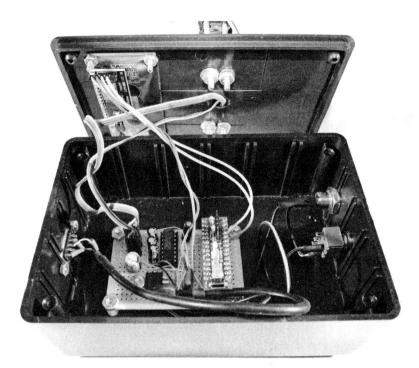

Figure 20.9 — Inside view showing the Ardweeny and support components.

Figure 20.10 — Close-up view of the Satellite Tracker's LCD display.

SDX\SDXParam.SQF file to change the settings manually. The first line in this file is the COM port number and the second line is the baud rate, which must be 9600 unless you change the baud rate setting in the sketch to match the PC baud rate setting.

Enhancement Ideas

When this project was originally designed, I was unaware that the Arduino Nano had an onboard voltage regulator. The Ardweeny could easily be replaced with a Nano, or even an Uno with a protoshield mounted inside the enclosure, and the external 5 V regulator portion of the circuit eliminated entirely. You could also replace the RS-232 interface and use the Nano or Uno's onboard USB port for the serial link to the software.

The sketch for this project was written for *SatPC32* and will not work with *Ham Radio Deluxe* (*HRD*) as it is currently coded. This is due to the fact that

Figure 20.11 — The finished Field Day Satellite Tracker

SatPC32 doesn't look for any response from the Ardweeny; it just sends the rotator control commands and assumes the controller executed them. *HRD* does a two-way handshake with the controller and expects a response to the rotator control commands. The sketch could be easily modified to respond back to *HRD* with the positioning information, allowing you to use the satellite tracking portion of *HRD* to control the Tracker. One thing to keep in mind, when *HRD* initially connects to the rotator controller, it toggles the DTR signal line on the USB interface. The Arduino Uno and Nano interpret this as a RESET command and resets the Arduino. As you will see in the full-scale rotator controller projects, adding a capacitor and a jumper block to the reset line will allow you to enable the auto-reset for loading your sketch and then disable it for use with *Ham Radio Deluxe*.

References

Atmel Corp — **www.atmel.com**
RadioShack — **www.radioshack.com**
SatPC32 software — **www.dk1tb.de**
Solarbotics — **www.solarbotics.com**
Yaesu — **www.yaesu.com**

CHAPTER 21

Azimuth/Elevation Rotator Controller

Yaesu G-5400B azimuth/elevation dual rotator controller.

Based on the success of the Field Day Satellite Tracker prototype, the next logical step is to build a full-scale Arduino controller for the actual Yaesu G5400/5500 series azimuth/elevation (az/el) rotator using either *SatPC32* or *Ham Radio Deluxe* (*HRD*). For automatic operation, the Yaesu G5400/5500 series rotator controllers are designed to work with the Yaesu GS-232 rotator computer interface. There are several different models of the GS-232 with slight differences among them. For this project, shown in **Figure 21.1**, we will be emulating the Yaesu GS-232A interface.

The Yaesu G5400/5500 series dual rotator controllers have a built in 8-pin DIN connector labeled EXTERNAL CONTROL (**Figure 21.2**). This connector provides an analog voltage output for the azimuth and elevation values, and relay control inputs for azimuth and elevation control. Ordinarily you would connect the Yaesu GS-232 rotator computer interface unit to this connector, but we're going to replace the GS-232 with an Arduino and have the Arduino emulate the GS-232 interface. We'll use the Arduino's USB port to interface with *SatPC32* or *Ham Radio Deluxe* on the PC side of things. We'll also be using the Arduino's onboard EEPROM to save our calibration settings. This way, the Azimuth/Elevation Rotator Controller will remember the calibration data even after a reset or power-off.

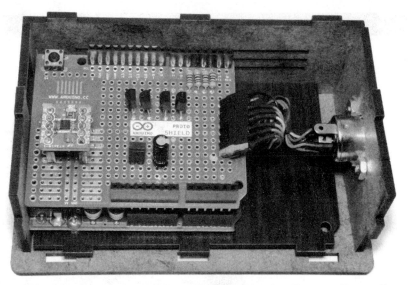

Figure 21.1 — Inside view of the Azimuth/Elevation Rotator Controller.

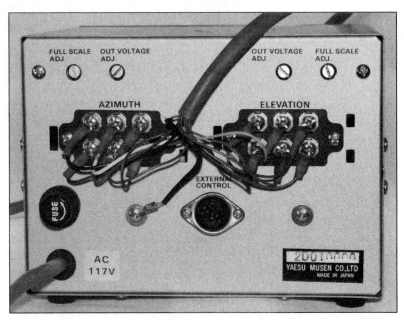

Figure 21.2 — Rear view of the Yaesu G-5400B rotator controller showing the External Control 8-pin DIN connector.

Figure 21.3 shows the block diagram for this project. The external hardware needed is minimal, comprising six resistors, four transistors, a capacitor, and a Texas Instruments ADS1115 4-channel, 16-bit I²C analog-to-digital (A/D) converter. You could use the Arduino's built in 10-bit A/D converter instead of the ADS1115, but you will not have the resolution that the extra six bits of the ADS1115 provides. The Arduino's A/D converter would only read three A/D counts per degree of rotation, versus the 148 A/D counts per degree with the ADS1115. Any noise on the positioning signals could cause positioning errors. Since we're tracking satellites, it's worth using the 16-bit A/D to get the extra accuracy.

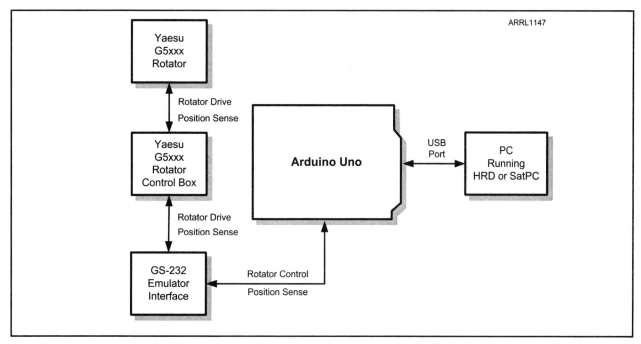

Figure 21.3 — Azimuth/Elevation Rotator Controller block diagram.

The Texas Instruments ADS1115 A/D Converter

Figure 21.4 — The Texas Instruments ADS1115 4-channel 16-bit analog-to-digital converter module.

The Texas Instruments ADS1115 (**Figure 21.4**) communicates with the Arduino using the I²C bus. The analog inputs can be configured as four single-ended inputs or two differential inputs, with the input gain program-selectable in six steps from 256 mV all the way up to 6.144 V. It is important to note that the 6.144 V gain setting is only used as a range on the A/D converter. At no time should a voltage higher than V_{DD} + 0.3 V (5.3 V) ever be used as an analog input to the A/D, otherwise you may damage the chip. The sampling rate is program-selectable in eight steps from 8 to 860 samples per second, along with a single-shot conversion mode. Because of the ADS1115's flexibility and ease of interfacing, it has become my A/D converter of choice when I need higher resolution than the Arduino's onboard 10-bit A/D converter.

Building the Hardware

Since the G5400/G5500 rotator controller already has most of the interface circuitry we'll need built-in, constructing the Arduino portion of the interface on the breadboard should go quickly. **Figure 21.5** shows the Fritzing diagram for this project. Note that we don't need relays to drive the rotator controller. The Yaesu rotator controller already has the control relays (with clamping diodes) built-in and connected to the external control connector, so all we need

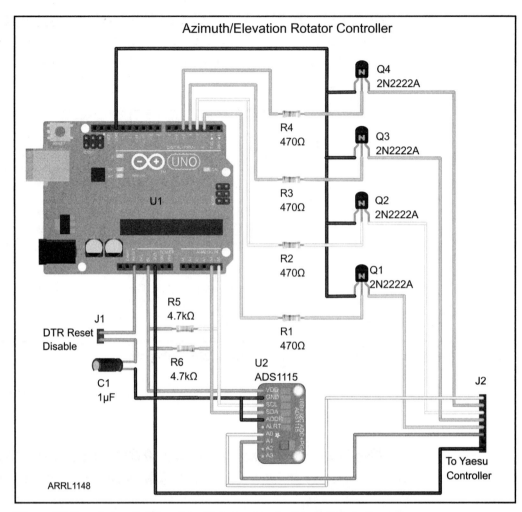

Figure 21.5 — Azimuth/Elevation Rotator Controller Fritzing diagram.

to do to drive the rotator is use a transistor for each rotator control pin. From your breadboard, you can build an 8-pin header-to-DIN-plug adapter cable to connect to the external control socket on the Yaesu controller. When you move the project to the protoshield, you can use an 8-pin header on the board to connect the DIN plug adapter cable you've already built. We'll power the Azimuth/Elevation Rotator Controller from the Arduino's USB connector that attaches to your PC, so we won't need an external source of power for this project.

Do not connect the Yaesu rotator controller to your breadboard circuit until you have calibrated the output voltages on the external control pins. First, use the Yaesu rotator controller to manually position the antenna to full scale for azimuth and elevation. Using a voltmeter, measure the voltage between the ground pin (pin 8) and elevation voltage pin (pin 1). Adjust the ELEVATION OUT VOLTAGE ADJ potentiometer on the back of the rotator controller for 5 V. Do the same for the azimuth voltage pin (pin 6) and the AZIMUTH OUT VOLTAGE ADJ potentiometer. Do not connect the DIN connector to your breadboard until you have completed this procedure, otherwise you could send a voltage higher than 5 V to the A/D converter and damage it. When you have completed the output voltage calibration, you should be able to rotate your azimuth and elevation rotators and see the output change from 0 to 5 V on each of the voltage output pins as you rotate the antenna from zero to full scale on the front panel meters.

The Sketch

When I initially began working with the TI ADS1115 A/D converter, the `ADS1115.h` and `I2Cdev.h` libraries did not handle the 16-bit data coming from the ADS1115 correctly. I had to modify these libraries to correct this issue. Be sure to use the `ADS1115.h` and `I2Cdev.h` libraries in **Appendix A** or download them from **www.w5obm.us/Arduino**.

Before you can load your sketch, you will need to be sure that the jumper that ties the 1 µF capacitor (C1) to the Reset pin is off. This capacitor is used to block the Reset command that *Ham Radio Deluxe* accidentally sends to the Arduino when it toggles the DTR signal line as it begins a connection attempt to the rotator controller. If you forget to short this jumper after loading your sketch, *Ham Radio Deluxe* will not be able to connect to the controller since the controller will be resetting while *HRD* is trying to connect to it. However, you do want to enable the DTR Reset when you load sketches, so be sure and remove the jumper before trying to load your sketch.

Next it's time to plan out the sketch. **Figure 21.6** shows the flowchart for the sketch and **Table 21.1** shows the actual Yaesu GS-232A commands that we will be implementing. This sketch will use many of the tools and techniques we've learned from the previous projects. For example, we'll be using the Texas Instruments ADS1115 to read the position voltages and we will be saving the position calibration values in the Arduino's onboard EEPROM. Using the Arduino's USB port, we will communicate with the satellite tracking software on the PC. To aid in troubleshooting, we will also implement a debug mode in the sketch, that when set, will return diagnostic information to the Arduino IDE's Serial Monitor.

Table 21.1
Azimuth/Elevation Rotator Controller Commands
(Subset of Yaesu GS-232A Commands)

A	Stop Azimuth Rotation
B	Return Current Elevation Value in Degrees (format +0eee)
C	Return Current Azimuth Value in Degrees (format +0aaa)
C2	Return Current Azimuth and Elevation (format +0aaa+0eee)
D	Rotate Elevation Down
E	Stop Elevation Rotation
F	Set Azimuth Full Scale Calibration
F2	Set Elevation Full Scale Calibration
L	Rotate Azimuth Counter-Clockwise
Maaa	Rotate Azimuth to aaa degrees
O	Set Azimuth Zero Calibration
O2	Set Elevation Zero Calibration
R	Rotate Azimuth Clockwise
S	Stop All Rotation
U	Rotate Elevation Up
Waaa eee	Rotate Azimuth to aaa degrees and Elevation to eee degrees

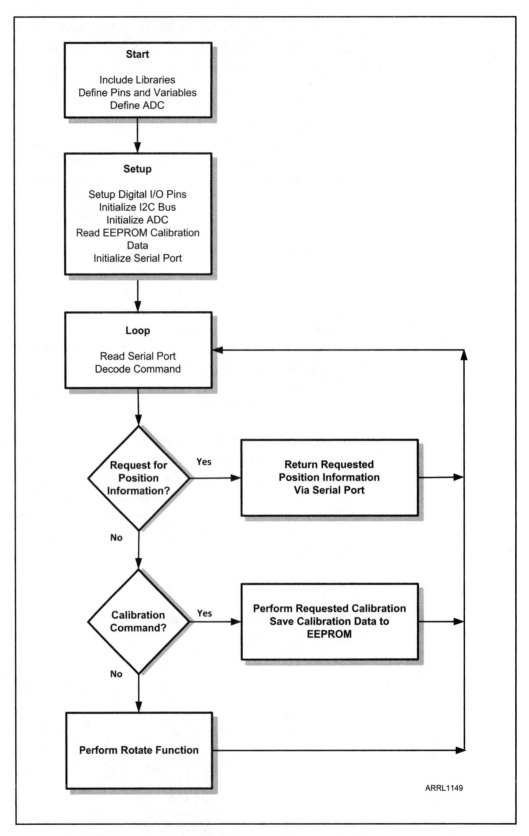

Figure 21.6 — Azimuth/Elevation Rotator Controller flowchart.

From the flowchart, you can see that in the main `loop()` we decode the command string from the tracking software on the PC into three basic groups of functions: status requests, calibration commands, and movement commands. By using this divide and conquer method, we break the sketch down into manageable chunks that can be independently created and debugged. The complete sketch and libraries can be found in **Appendix A** or online at **www.w5obm.us/Arduino**.

Starting out in the sketch, we'll define the libraries and the ADS1115 object. We also include a definition for the debug mode. When this definition is set to a "1", debug information will be sent to the Arduino IDE's Serial Monitor.

```
#define debug_mode 0  // Set to 1 for debug data on Serial Port

#include <Wire.h> // Include the I2C Communication Library
#include <EEPROM.h>  // Include the EEPROM Library

// Include the ADS1115.h Library (Library Updated to fix errors)
#include "ADS1115.h"

// Include I2Cdev.h Library (Library Updated to fix errors)
#include "I2Cdev.h"


ADS1115 adc; // Define the ADS1115 as adc
```

Next, we'll define the Arduino digital I/O pins used to control the antenna rotation and the baud rate for the serial port used to interface with the tracking software on your PC. The baud rate in the sketch must match the baud rate you have set in the tracking software on your PC.

```
#define rotate_up 3   // Define Rotate Up as Pin 3
#define rotate_down 2  // Define Rotate Down as Pin 2
#define rotate_left 4  // Define Rotate Left as Pin 4
#define rotate_right 5  // Define Rotate Right as Pin 5

#define BAUD_RATE 9600  // Set the Serial Port Baud rate to 9600
```

Since we will be storing the positioning calibration data in EEPROM, we will need to define the format we will be using to read and store the EEPROM data.

```
// EEPROM ID to validate EEPROM data location
#define EEPROM_ID_BYTE 1
 // EEPROM ID Value
#define EEPROM_ID 55
// Azimuth Zero Calibration EEPROM location
#define EEPROM_AZ_CAL_0 2
// Azimuth Max Calibration Data EEPROM location
```

```
#define EEPROM_AZ_CAL_MAX 4
// Elevation Zero Calibration Data EEPROM location
#define EEPROM_EL_CAL_0 6
// Elevation Max Calibration Data EEPROM location
#define EEPROM_EL_CAL_MAX 8
```

In case there is no calibration data stored in the EEPROM, default values are defined, along with a "tolerance" value for the rotator positioning. The tolerance is a value in degrees that the controller can use to determine that the current rotator positioning is within the tolerance range to the desired position. The tolerance value is designed for use when you have excessive noise on the positioning signals and want to prevent rotation when the desired position and the actual position are calculated to be a few degrees apart from each other. In practical use, a tolerance of 0 works well in my shack.

```
// Preset the Azimuth Zero Calibration Point to 0
#define AZ_CAL_0_DEFAULT 0
// Preset the Azimuth Max Calibration Point to 27000
#define AZ_CAL_MAX_DEFAULT 27000
// Preset the Elevation Zero Calibration Point to 0
#define EL_CAL_0_DEFAULT 0
// Preset the Elevation Max Calibration Point to 27000
#define EL_CAL_MAX_DEFAULT 27000

#define AZ_Tolerance 0   // Set the Azimuth Accuracy Tolerance
#define EL_Tolerance 0   // Set the Elevation Accuracy Tolerance
```

Next, we'll define all the variables used in the sketch:

```
byte inByte = 0;   // incoming serial byte
byte serial_buffer[50];   // incoming serial byte buffer

// The index pointer variable for the Serial buffer
int serial_buffer_index = 0;

int set_AZ;   // Azimuth set value
int set_EL;   // Elevation set value
int current_AZ;   // Current Azimuth raw value
int current_EL;   // Current Elevation raw value
String Serial_Send_Data; // Data to send to Serial Port
int AZ_0;   // Azimuth Zero Value from EEPROM
int AZ_MAX; // Azimuth Max Value from EEPROM
int EL_0;   // Elevation 0 Value from EEPROM
int EL_MAX;   //Elevation Max Value from EEPROM
int AZ_Degrees; // mapped AZ ADC value to Degrees
int EL_Degrees; // mapped EL ADC value to Degrees
String Requested_AZ; // RS232 Requested Azimuth - M and short W command
String Requested_EL; //RS232 Requested Azimuth and Elevation - Full W command
```

```
int AZ_To; // Requested AZ Move
int EL_To; // Requested EL Move
int AZ_Distance; // Distance to move AZ
int EL_Distance; // Distance to move EL
```

In the `setup()` loop, we'll initialize the digital I/O pins and make sure that all the position control outputs are off:

```
pinMode(rotate_up, OUTPUT);    // Define the Control Pins as Outputs
pinMode(rotate_down, OUTPUT);
pinMode(rotate_left, OUTPUT);
pinMode(rotate_right, OUTPUT);

digitalWrite(rotate_up, LOW);  // Turn off all the Control Pins
digitalWrite(rotate_down, LOW);
digitalWrite(rotate_left, LOW);
digitalWrite(rotate_right, LOW);
```

Now we'll initialize the serial port, the I²C bus, and the ADS1115 A/D converter:

```
Serial.begin(BAUD_RATE); // initialize serial communication

Wire.begin();  // join I2C bus

adc.initialize(); // initialize ADS1115 16 bit A/D chip

Wire.beginTransmission(0x48); // Begin direct ADC communication

//  Connect to adc and send two bytes - Set Config Reg to all Ones
// We do this so when we read the ADC we can be sure we're
// communicating correctly
Wire.write(0x1);
Wire.write(0x7F);  //  MSB
Wire.write(0xFF);  //  LSB
Wire.endTransmission();  // End the direct ADC Communication
```

As the final step of initializing the A/D converter, we'll set it to free-running single-ended mode at 475 samples per second, with a range of 0 to 6.144 V.

```
// Set the ADC to free running conversion mode
adc.setMode(ADS1115_MODE_CONTINUOUS);

// set the ADC gain to 6.144 Volt range, .0001875 Volts/step
adc.setGain(ADS1115_PGA_6P144);

// set ADC sample rate to 475 samples per second
adc.setRate(ADS1115_RATE_475);
```

```
// Set the ADC to AN0+ Vs ground Mode (single-ended)
adc.setMultiplexer(ADS1115_MUX_P0_NG);
```

As the last step in the `setup()` loop, the move flag variables are set to -1 to indicate that there is no active rotation command, and the calibration values (if any) are loaded from EEPROM.

```
Set_AZ = -1;  // Preset the Azimuth and Elevation Move Variables
set_EL = -1;

// Read the Azimuth and Elevation Calibration Values from EEPROM
read_eeprom_cal_data();
```

The main `loop()` for this sketch is rather unique. It consists of only two statements. The divide and conquer design strategy for this project resulted in the majority of the work being handled by functions. All the main loop does is continually check for incoming commands on the serial port, decode and execute any commands it receives, and handle any commands involving move commands. Each decoded command will call a function to perform the action called for by the incoming command.

```
Check_serial();  // Check the Serial Port for Data
check_move();    // Check to see if executing move command
```

The majority of this project is controlled by the various functions. As part of the `setup()` loop, the `read_eeprom_cal_data()` function is called to load the saved calibration data into the sketch variables. This function checks to see if there is a valid EEPROM ID byte and loads the EEPROM data into the sketch variables if the ID is valid. If the EEPROM ID is not valid, the sketch calibration variables are loaded with the defined default values and saved to EEPROM. If the debug mode is enabled, the IDE's Serial Monitor will display the values read from the EEPROM.

```
// Function to Read the Azimuth and Elevation Calibration Data
void read_eeprom_cal_data()
{
  // Verify the EEPROM has valid data
  if (EEPROM.read(EEPROM_ID_BYTE) == EEPROM_ID)
  {
    if (debug_mode) // If in Debug Mode Print the Calibration Values
    {
      Serial.println("Read EEPROM Calibration Data Valid ID");
      Serial.println((EEPROM.read(EEPROM_AZ_CAL_0) * 256) + EEPROM.read(EEPROM_AZ_CAL_0 + 1),DEC);
      Serial.println((EEPROM.read(EEPROM_AZ_CAL_MAX) * 256) + EEPROM.read(EEPROM_AZ_CAL_MAX + 1),DEC);
      Serial.println((EEPROM.read(EEPROM_EL_CAL_0) * 256) + EEPROM.read(EEPROM_EL_CAL_0 +1 ),DEC);
      Serial.println((EEPROM.read(EEPROM_EL_CAL_MAX) *256) + EEPROM.read(EEPROM_
```

```
  EL_CAL_MAX + 1),DEC);
      }
      // Read the Azimuth Zero Calibration Value from EEPROM
      AZ_0 = (EEPROM.read(EEPROM_AZ_CAL_0)*256) + EEPROM.read(EEPROM_AZ_CAL_0 + 1);
      // Read the Azimuth Maximum Calibration Value from EEPROM
      AZ_MAX = (EEPROM.read(EEPROM_AZ_CAL_MAX)*256) + EEPROM.read(EEPROM_AZ_CAL_MAX + 1);
      // Read the Elevation Zero Calibration Value from EEPROM
      EL_0 = (EEPROM.read(EEPROM_EL_CAL_0)*256) + EEPROM.read(EEPROM_EL_CAL_0 + 1);
      // Read the Elevation Maximum Calibration Value from EEPROM
      EL_MAX = (EEPROM.read(EEPROM_EL_CAL_MAX)*256) + EEPROM.read(EEPROM_EL_CAL_MAX + 1);

  } else {  // initialize eeprom to default values
  if (debug_mode)
  {
    Serial.println("Read EEPROM Calibration Data Invalid ID - setting to defaults");
  }
  AZ_0 = AZ_CAL_0_DEFAULT;   // Set the Calibration To Default Values
  AZ_MAX = AZ_CAL_MAX_DEFAULT;
  EL_0 = EL_CAL_0_DEFAULT;
  EL_MAX = EL_CAL_MAX_DEFAULT;
  write_eeprom_cal_data();   // Write the Default Values to EEPROM
  }
}
```

The `write_eeprom_cal_data()` function is called when data is to be saved to the EEPROM, either as part of the initial EEPROM setup, or when a calibration command is received via the USB serial port. The 16 bit integer values are broken into high and low order bytes and written to the EEPROM one byte at a time.

```
// Function to Write the Calibration Values to EEPROM
void write_eeprom_cal_data()
{
  if (debug_mode)
  {
    Serial.println("Writing EEPROM Calibration Data");
  }

  EEPROM.write(EEPROM_ID_BYTE,EEPROM_ID);  //   Write the EEPROM ID

  // Write the Azimuth Zero Calibration High Order Byte
  EEPROM.write(EEPROM_AZ_CAL_0,highByte(AZ_0));
```

```cpp
    // Write the Azimuth Zero Calibration Low Order Byte
    EEPROM.write(EEPROM_AZ_CAL_0 + 1,lowByte(AZ_0));

    // Write the Azimuth Max Calibration High Order Byte
    EEPROM.write(EEPROM_AZ_CAL_MAX,highByte(AZ_MAX));

    // Write the Azimuth Max Calibration Low Order Byte
    EEPROM.write(EEPROM_AZ_CAL_MAX + 1,lowByte(AZ_MAX));

    // Write the Elevation Zero Calibration High Order Byte
    EEPROM.write(EEPROM_EL_CAL_0,highByte(EL_0));

    // Write the Elevation Zero Calibration Low Order Byte
    EEPROM.write(EEPROM_EL_CAL_0 + 1,lowByte(EL_0));

    // Write the Elevation Max Calibration High Order Byte
    EEPROM.write(EEPROM_EL_CAL_MAX,highByte(EL_MAX));

    // Write the Elevation Max Calibration Low Order Byte
    EEPROM.write(EEPROM_EL_CAL_MAX + 1,lowByte(EL_MAX));
}
```

The check_serial() function is where the serial commands from the PC tracking software are received and decoded. The function echoes all incoming characters back to the tracking software. This is required for the *SatPC32* rotator control program to function. If you are using *Ham Radio Deluxe* to control the rotator, you may need to comment out the portion of the sketch that echoes the command back to the PC. As each character is received, it is added to the serial buffer until a carriage return is received.

```cpp
// Function to check for data on the Serial port
void check_serial()
{
  if (Serial.available() > 0) // Get the Serial Data if available
  {
    inByte = Serial.read();   // Get the Serial Data

   // You may need to comment out the following line if your PC software
   // will not communicate properly with the controller
   // SatPC32 wants the command echoed, Ham Radio Deluxe does not
    Serial.print(char(inByte));  // Echo back to the PC

    if (inByte == 10)   // ignore Line Feeds
    {
    return;
    }
    if (inByte !=13)    // Add to buffer if not CR
    {
```

```
    serial_buffer[serial_buffer_index] = inByte;

    if (debug_mode) // Print the Character received if in Debug mode
    {
      Serial.print("Received = ");
      Serial.println(serial_buffer[serial_buffer_index]);
    }
    serial_buffer_index++;  // Increment the Serial Buffer pointer

  } else {  // It's a Carriage Return, execute command

  // If first character of command is lowercase, convert to uppercase
  if ((serial_buffer[0] > 96) && (serial_buffer[0] < 123))
  {
    serial_buffer[0] = serial_buffer[0] - 32;
  }
```

When a carriage return is received, the first character of the command is converted to uppercase as necessary, and the function is decoded and executed. A `switch...case()` statement is used to decode the various commands, and each command is executed by a function within the `switch...case()` statement. If the debug mode is enabled, additional diagnostic information will be sent to the Arduino IDE's Serial Monitor. The `serial_buffer_index` variable indicates how many characters were received for the decoded command, and is used to decode commands that have multiple options.

```
// Decode first character of command
switch (serial_buffer[0])
{

  case 65:  // A Command - Stop the Azimuth Rotation
    if (debug_mode) {Serial.println("A Command Received");}
    az_rotate_stop(); // Call the Azimuth Rotate Stop function
    break;

  case 66:  // B Command - Send the Current Elevation to the PC
    if (debug_mode) {Serial.println("B Command Received");}
    send_current_el();  // Call the Send Current Elevation Function
    break;
```

The "C" command will send the current azimuth in degrees back to the tracking software on the PC. However, this command may actually be a "C2" command, which needs to send the current azimuth and elevation back to the tracking software. The length of the command string as determined by the `serial_buffer_index` variable is used to decide what information needs to be sent back to the tracking software.

```
  Case 67:          // C - return current azimuth
    if (debug_mode)    // Return the Buffer Index Pointer in Debug Mode
    {
      Serial.println("C Command Received");
      Serial.println(serial_buffer_index);
    }

    // check for C2 command
    if ((serial_buffer_index == 2) & (serial_buffer[1] == 50))
    {
      if (debug_mode)
      {
        Serial.println("C2 Command Received");
      }
      send_current_azel();  // Return Azimuth and Elevation if C2 Command
    } else {
      send_current_az();   // Return Azimuth if C Command
    }
    break;

  case 68:  // D - Rotate Elevation Down Command
    if (debug_mode)
    {
      Serial.println("D Command Received");
    }
    rotate_el_down();  // Call the Rotate Elevation Down Function
    break;

  case 69:  // E - Stop the Elevation Rotation
    if (debug_mode)
    {
      Serial.println("E Command Received");
    }
    el_rotate_stop();  // Call the Elevation Rotation Stop Function
    break;
```

The "F" command is another command that can have a secondary function. The "F" command by itself will read and save the azimuth full scale calibration value, while the "F2" command will read and save the elevation full scale value.

```
  Case 70:  // F - Set the Max Calibration
    if (debug_mode)
    {
      Serial.println("F Command Received");
      Serial.println(serial_buffer_index);
    }
    // Check for F2 Command
    if ((serial_buffer_index == 2) & (serial_buffer[1] == 50))
```

```
    {
      if (debug_mode)
      {
        Serial.println("F2 Command Received");
      }
      set_max_el_cal();  // F2 - Set the Max Elevation Calibration
    } else {
      set_max_az_cal();  // F - Set the Max Azimuth Calibration
    }
    break;

    case 76:  // L - Rotate Azimuth CCW
    if (debug_mode)
    {
      Serial.println("L Command Received");
    }
    rotate_az_ccw();  // Call the Rotate Azimuth CCW Function
    break;

  case 77:  // M - Rotate to Set Point
    if (debug_mode)
    {
      Serial.println("M Command Received");
    }
    rotate_to();  // Call the Rotate to Set Point Command
    break;
```

> The "O" command is used for the zero calibration settings. An "O" command by itself will save the azimuth zero calibration value, while an "O2" command is used to save the elevation zero calibration value.

```
Case 79:  // O - Set Zero Calibration
  if (debug_mode)
  {
    Serial.println("O Command Received");
    Serial.println(serial_buffer_index);
  }
  // Check for O2 Command
  if ((serial_buffer_index == 2) & (serial_buffer[1] == 50))
  {
    if (debug_mode)
    {
      Serial.println("O2 Command Received");
    }
    set_0_el_cal();  // O2 - Set the Elevation Zero Calibration
  } else {
    set_0_az_cal();  // O - Set the Azimuth Zero Calibration
  }
  break;
```

Azimuth/Elevation Rotator Controller

```
  case 82:   // R - Rotate Azimuth CW
    if (debug_mode)
    {
      Serial.println("R Command Received");
    }
    rotate_az_cw();  // Call the Rotate Azimuth CW Function
    break;

case 83:   // S - Stop All Rotation
  if (debug_mode)
  {
    Serial.println("S Command Received");
  }
  az_rotate_stop();  // Call the Stop Azimith Rotation Function
  el_rotate_stop();  // Call the Stop Elevation Rotation Function
  break;

case 85:   // U - Rotate Elevation Up
  if (debug_mode)
  {
    Serial.println("U Command Received");
  }
  rotate_el_up();  // Call the Rotate Elevation Up Function
  break;

case 87:   // W - Rotate Azimuth and Elevation to Set Point
  if (debug_mode)
  {
    Serial.println("W Command Received");
  }
  // Call the Rotate Azimuth and Elevation to Set Point Function
  rotate_az_el_to();
  break;

}
```

When the command has been decoded and executed, the incoming serial character buffer is cleared and the buffer index is set to 0. The function will then return to the main `loop()` to wait for the next command.

```
// Clear the Serial Buffer and Reset the Buffer Index Pointer
serial_buffer_index = 0;
serial_buffer[0] = 0;
```

The `send_current_az()` function is used to build and send the response to the "C" command sent by the tracking software on the PC. The tracking software expects the response to be in the format of "+0aaa", where aaa is the azimuth value in degrees. This function will read the A/D converter and map the value to degrees based on the current azimuth calibration values. A string is then built using the Yaesu GS-232A format that the tracking software is expecting, and then sent back to the tracking software via the USB serial port. Because the G5400/5500 rotators have 0°/360° at midscale, the data has to be adjusted to return the correct azimuth value.

```
// Send the Current Azimuth Function
void send_current_az()
{
  read_adc();  // Read the ADC

  // Map Azimuth to degrees
  if (debug_mode)
  {
    Serial.println(current_AZ);
  }

  // Map the Current Azimuth to Degrees
  AZ_Degrees = map(current_AZ, AZ_0, AZ_MAX, 0, 360);

  // Correction Since Azimuth Reading starts at Meter Center Point
  if (AZ_Degrees > 180)
  {
    AZ_Degrees = AZ_Degrees - 180;
  } else {
    AZ_Degrees = AZ_Degrees + 180;
  }
  if (debug_mode)
  {
    Serial.println(AZ_Degrees);
  }
  // Send it back via serial
  Serial_Send_Data = "";
  if (AZ_Degrees < 100)  // pad with 0's if needed
  {
    Serial_Send_Data = "0";
  }
  if (AZ_Degrees < 10)
  {
    Serial_Send_Data = "00";
  }
  // Send the Azimuth in Degrees
  Serial_Send_Data = "+0" + Serial_Send_Data + String(AZ_Degrees);
  Serial.println(Serial_Send_Data);  // Return value via RS-232 port
}
```

The `send_current_azel()` function is used to build and send the response to the "C2" command sent by the tracking software back to the PC. The tracking software expects the response to be in the format of "+0aaa+0eee", where aaa is the azimuth value, and eee is the elevation value. This function will read the A/D converter and map the values to degrees based on the current azimuth and elevation calibration values. A string is then built using the Yaesu GS-232A format that the tracking software is expecting, and then sent back to the tracking software via the USB serial port. As with the "C" command, the azimuth data has to be adjusted to return the correct azimuth value.

```
// Function to Send the Current Azimuth and Elevation
void send_current_azel()
{
  read_adc();    // Read the ADC

  // Map Azimuth to degrees
  if (debug_mode)
  {
    Serial.println(current_AZ);
  }
  // Map the Current Azimuth to Degrees
  AZ_Degrees = map(current_AZ, AZ_0, AZ_MAX, 0, 360);

  // Correction Since Azimuth Reading starts at Meter Center Point
  if (AZ_Degrees > 180)
  {
    AZ_Degrees = AZ_Degrees - 180;
  } else {
    AZ_Degrees = AZ_Degrees + 180;
  }

  // Map Elevation to degrees
  if (debug_mode)
  {
    Serial.println(current_EL);
  }
  // Map the Elevation to Degrees
  EL_Degrees = map(current_EL, EL_0, EL_MAX, 0, 180);
  if (debug_mode)
  {
    Serial.println(EL_Degrees);
    Serial.println(AZ_Degrees);
  }
  // Send it back via serial
  Serial_Send_Data = "";
  if (AZ_Degrees < 100)   // pad with 0's if needed
  {
    Serial_Send_Data = "0";
```

```
  }
  if (AZ_Degrees < 10)
  {
    Serial_Send_Data = "00";
  }
  // Send the Azimuth part of the string
  Serial_Send_Data = "+0" + Serial_Send_Data + String(AZ_Degrees) + "+0";
  if (EL_Degrees < 100)  // pad with 0's if needed
  {
    Serial_Send_Data = Serial_Send_Data + "0";
  }
  if (EL_Degrees < 10)
  {
    Serial_Send_Data = Serial_Send_Data+ "0";
  }
  // Send the Elevation Part of the String
  Serial_Send_Data = Serial_Send_Data + String(EL_Degrees);
  Serial.println(Serial_Send_Data);   // Return value via Serial port
}
```

The `set_max_az_cal()` function is used to save the azimuth full scale value to the Arduino's onboard EEPROM:

```
// Set the Max Azimuth Calibration Function
void set_max_az_cal()
{
  if (debug_mode)
  {
    Serial.println("Calibrate Max AZ Function");
  }
  read_adc();   // Read the ADC

  // save current az and el values to EEPROM - Zero Calibration
  if (debug_mode)
  {
    Serial.println(current_AZ);
  }
  // Set the Azimuth Maximum Calibration to Current Azimuth Reading
  AZ_MAX = current_AZ;
  write_eeprom_cal_data();   // Write the Calibration Data to EEPROM
  if (debug_mode)
  {
    Serial.println("Max Azimuth Calibration Complete");
  }
}
```

The `set_max_el_cal()` function is used to save the elevation full scale value to the Arduino's onboard EEPROM:

```
// Set the Max Elevation Calibration Function
void set_max_el_cal()
{
  if (debug_mode)
  {
    Serial.println("Calibrate EL Max Function");
  }
  read_adc();   // Read the ADC

  // save current Azimuth and Elevation values to EEPROM - Zero Calibration
  if (debug_mode)
  {
    Serial.println(current_EL);
  }
  // Set the Elevation Max Calibration to the Current Elevation Reading
  EL_MAX = current_EL;

  write_eeprom_cal_data();   // Write the Calibration Data to EEPROM
  if (debug_mode)
  {
    Serial.println("Max Elevation Calibration Complete");
  }
}
```

The `rotate_az_ccw()`, `rotate_az_cw()`, `rotate_el_up()`, and `rotate_el_down()` functions control the relays inside the Yaesu G5400/5500 rotator controller to perform the actual rotation functions. The `az_rotate_stop()` and `el_rotate_stop()` are used to turn off the relays and stop rotation.

```
// Function to Rotate Azimuth CCW
void rotate_az_ccw()
{
  digitalWrite(rotate_left, HIGH);   // Set the Rotate Left Pin High
  digitalWrite(rotate_right, LOW);   // Make sure the Rotate Right Pin is Low
}

// Function to Rotate Azimuth CW
void rotate_az_cw()
{
  digitalWrite(rotate_right, HIGH);   // Set the Rotate Right Pin High
  digitalWrite(rotate_left, LOW);     // Make sure the Rotate Left Pin Low
}

// Function to Rotate Elevation Up
void rotate_el_up()
{
  digitalWrite(rotate_up, HIGH);   // Set the Rotate Up Pin High
```

```
  digitalWrite(rotate_down, LOW);  // Make sure the Rotate Down Pin is Low
}

// Function to Rotate Elevation Up
void rotate_el_down()
{
  digitalWrite(rotate_down, HIGH);  // Set the Rotate Down Pin High
  digitalWrite(rotate_up, LOW);     // Make sure the Rotate Up Pin is Low
}

// Function to Stop Azimuth Rotation
void az_rotate_stop()
{
  digitalWrite(rotate_right, LOW);  // Turn off the Rotate Right Pin
  digitalWrite(rotate_left, LOW);   // Turn off the Rotate Left Pin
}

// Function to Stop Elevation Rotation
void el_rotate_stop()
{
  digitalWrite(rotate_up, LOW);    // Turn off the Rotate Up Pin
  digitalWrite(rotate_down, LOW);  // Turn off the Rotate Down Pin
}
```

The `rotate_to()` function is used to implement the "M" command. This command is in the format of Maaa, where aaa is the desired azimuth value in degrees. This function will read the desired azimuth from the "M" command, compare the requested value to the reading from the A/D, and rotate the azimuth rotator in the proper direction. While the rotator is moving to the desired position, this function will allow processing to return to the main `loop()` and check for additional commands, such as a "Stop" command. The status and control of this move command is handled in the main `loop()` by the `check_move()` function.

```
// Function to Rotate to Set Point
void rotate_to()
{
  if (debug_mode)
  {
    Serial.println("M Command - Rotate Azimuth To Function");
  }
  // Decode Command - Format Mxxx where xxx = Degrees to Move to
  if (debug_mode)
  {
    Serial.println(serial_buffer_index);
  }
  if (serial_buffer_index == 4)  // Verify the Command is the proper length
  {
```

```
  if (debug_mode)
  {
    Serial.println("Value in [1] to [3]?");
  }
  // Decode the Azimuth Value
  Requested_AZ = (String(char(serial_buffer[1])) + String(char(serial_buffer[2])) + String(char(serial_buffer[3]))) ;

  AZ_To = (Requested_AZ.toInt()); // AZ Degrees to Move to as integer
  if (AZ_To <0) // Make sure we don't go below 0 degrees
  {
    AZ_To = 0;
  }
  if (AZ_To >360) // Make sure we don't go over 360 degrees
  {
    AZ_To = 360;
  }
  if (AZ_To > 180) // Adjust for Meter starting at Center
  {
    AZ_To = AZ_To - 180;
  } else {
    AZ_To = AZ_To + 180;
  }
  if (debug_mode)
  {
    Serial.println(Requested_AZ);
    Serial.println(AZ_To);
  }

  // set the move flag and start
  read_adc();   // Read the ADC

  // Map it to degrees
  if (debug_mode)
  {
    Serial.println(current_AZ);
  }
  // Map the Azimuth Value to Degrees
  AZ_Degrees = map(current_AZ, AZ_0, AZ_MAX, 0, 360);
  if (debug_mode)
  {
    Serial.println(AZ_Degrees);
  }
  AZ_Distance = AZ_To - AZ_Degrees;   // Figure out far we have to move
  set_AZ = AZ_To;
  // No move needed if we're within the Tolerance Range
  if (abs(AZ_Distance) <= AZ_Tolerance)
  {
```

```
      az_rotate_stop();  // Stop the Azimuth Rotation
      set_AZ = -1;  // Turn off the Move Command
    } else {  // Move Azimuth - figure out which way
      if (AZ_Distance > 0)   //We need to move CCW
      {
        rotate_az_ccw();  // If the distance is positive, move CCW
      } else {
        rotate_az_cw();  // Otherwise, move clockwise
      }
    }
  }
}
```

The `send_current_el()` function prepares and returns the response to the "B" command. The response is in the format "+0eee", where eee is the current elevation value in degrees.

```
// Function to Send the Current Elevation
void send_current_el()
{
  read_adc();  // Read the ADC
  // Map it to degrees
  if (debug_mode)
  {
    Serial.println(current_EL);
  }
  // Map the Elevation Value to Degrees
  EL_Degrees = map(current_EL, EL_0, EL_MAX, 0, 180);
  if (debug_mode)
  {
    Serial.println(EL_Degrees);
    Serial.println(AZ_Degrees);
  }
  // Send it back via serial
  Serial_Send_Data = "";
  if (EL_Degrees < 100)  // pad with 0's if needed
  {
    Serial_Send_Data = "0";
  }
  if (EL_Degrees < 10)
  {
    Serial_Send_Data = "00";
  }
  // Send the Elevation String
  Serial_Send_Data = "+0" + Serial_Send_Data + String(EL_Degrees);
  Serial.println(Serial_Send_Data);  // Return value via RS-232 port
}
```

The `rotate_az_el_to()` function is similar to the `rotate_to()` function, except it performs azimuth and elevation rotation simultaneously in response to the "W" command. The "W" command is in the format "Waaa eee", where aaa is the desired azimuth, and eee is the desired elevation. This function will read the desired azimuth and elevation from the "W" command, compare the requested values to the readings from the A/D, and rotate the azimuth and elevation rotators in the proper direction. While the rotators are moving to the desired position, this function will allow processing to return to the main `loop()` and check for additional commands, such as a "Stop" command. The status and control of this move command is handled in the main `loop()` by the `check_move()` function. Since the azimuth and elevation rotations are handled separately, both axis of rotation can occur simultaneously, and each will stop independently when the desired position is reached.

```
// Rotate Azimuth and Elevation to Set Point Function
void rotate_az_el_to()
{
  if (debug_mode)
  {
    Serial.println("W Command -  Rotate Azimuth and Elevation To Function");
  }
  // Decode Command - Format Waaa eee where aaa = Azimuth
  // to Move to and eee = Elevation to Move to
  if (debug_mode)
  {
    Serial.println(serial_buffer_index);
  }
  if (serial_buffer_index == 8)  // Verify the command is the proper length
  {
    if (debug_mode)
    {
      Serial.println("Value in [1] to [3]?");
    }
  // Decode the Azimuth portion of the command
    Requested_AZ = (String(char(serial_buffer[1])) + String(char(serial_buffer[2])) + String(char(serial_buffer[3]))) ;

    AZ_To = (Requested_AZ.toInt()); // AZ Degrees to Move to as integer
    if (AZ_To <0) // Don't allow moving below zero
    {
      AZ_To = 0;
    }
    if (AZ_To >360) // Don't allow moving above 360
    {
      AZ_To = 360;
    }
    if (AZ_To >180) // Adjust for Azimuth starting at Center
    {
```

```
      AZ_To = AZ_To - 180;
    } else {
      AZ_To = AZ_To + 180;
    }

    if (debug_mode)
    {
      Serial.println(Requested_AZ);
      Serial.println(AZ_To);
      Serial.println("Value in [5] to [7]?");
    }
    // Decode the Elevation portion of the command
    Requested_EL = (String(char(serial_buffer[5])) + String(char(serial_
buffer[6])) + String(char(serial_buffer[7]))) ;

    EL_To = (Requested_EL.toInt()); // EL Degrees to Move to as integer
    if (EL_To <0) // Don't allow moving below zero
    {
      EL_To = 0;
    }
    if (EL_To >180) // Don't allow moving above 180
    {
      EL_To = 180;
    }
    if (debug_mode)
    {
      Serial.println(Requested_EL);
      Serial.println(EL_To);
    }

    // set the move flag and start
    read_adc();   // Read the ADC
    // Map it to degrees
    if (debug_mode)
    {
      Serial.println(current_AZ);
    }
    // Map the Azimuth Value to Degrees
    AZ_Degrees = map(current_AZ, AZ_0, AZ_MAX, 0, 360);
    if (debug_mode)
    {
      Serial.println(AZ_Degrees);
    }
    AZ_Distance = AZ_To - AZ_Degrees;  // Figure how far to move Azimuth
    set_AZ = AZ_To;
    // Map the Elevation Value to Degrees
    EL_Degrees = map(current_EL, EL_0, EL_MAX, 0, 180);
    if (debug_mode)
```

```
    {
      Serial.println(EL_Degrees);
    }
    EL_Distance = EL_To - EL_Degrees;  // Figure how far to move Elevation
    set_EL = EL_To;
    // Set Azimuth
    // No move needed if we're within tolerance range
    if (abs(AZ_Distance) <= AZ_Tolerance)
    {
      az_rotate_stop();   // Stop the Azimuth Rotation
      set_AZ = -1;   // Turn off the Azimuth Move Command
    } else {   // Move Azimuth - figure out which way
      if (AZ_Distance > 0)    //We need to move CW
      {
        rotate_az_cw();   // Rotate CW if positive
      } else {
        rotate_az_ccw();  // Rotate CCW if negative
      }
    }
    // Set Elevation
    // No move needed if we're within tolerance range
    if (abs(EL_Distance) <= EL_Tolerance)
    {
      el_rotate_stop();   // Stop the Elevation Rotation
      set_EL = -1;   // Turn off the Elevation Move Command
    } else {   // Move Elevation - figure out which way
      if (EL_Distance > 0)    //We need to move CW
      {
        rotate_el_up();   // Rotate Up if positive
      } else {
        rotate_el_down();   // Rotate Down if negative
      }
    }
  }
}
```

The `set_0_az_cal()` function is used implement the "O" command and saves the azimuth zero calibration value to the Arduino's EEPROM:

```
// Set Azimuth Zero Calibration
void set_0_az_cal()
{
  if (debug_mode)
  {
    Serial.println("Calibrate Zero Function");
  }

  read_adc();   // Read the ADC
```

```
  // save current az and el values to EEPROM - Zero Calibration
  if (debug_mode)
  {
    Serial.println(current_EL);
    Serial.println(current_AZ);
  }
  AZ_0 = current_AZ;  // Set the Azimuth Zero Calibration to current position
  write_eeprom_cal_data();  // Write the Calibration Data to EEPROM
  if (debug_mode)
  {
    Serial.println("Zero Azimuth Calibration Complete");
  }
}
```

The `set_0_el_cal()` function is used implement the "O2" command and saves the elevation zero calibration value to the Arduino's EEPROM:

```
//  Set the Elevation Zero Calibration
void set_0_el_cal()
{
  if (debug_mode)
  {
    Serial.println("Calibrate Zero Function");
  }

  read_adc();  // Read the ADC

  // save current az and el values to EEPROM - Zero Calibration
  if (debug_mode)
  {
    Serial.println(current_EL);
    Serial.println(current_AZ);
  }
  // Set the Elevation Zero Calibration to current position
  EL_0 = current_EL;
  write_eeprom_cal_data();  // Write the Calibration Data to EEPROM
  if (debug_mode)
  {
    Serial.println("Zero Elevation Calibration Complete");
  }
}
```

The `read_adc()` function is used to read the analog azimuth and elevation voltages coming from the rotator controller and converts these values to raw, uncalibrated, 16-bit digital values using the ADS1115 A/D. The values are stored in the global variables `current_EL` and `current_AZ`, and used by the other functions to determine the current azimuth and elevation of the rotators.

```
// Function to read the ADC
void read_adc()
{
 if (debug_mode)
 {
   Serial.println("Read ADC Function  ");
 }

  int RotorValue;  // Variable to store the rotor value
  adc.setRate(ADS1115_RATE_475); // Set the ADC rate to 475 samples/sec
  adc.setGain(ADS1115_PGA_6P144); // Set the ADC gain to 6.144V
  // Set the ADC to Channel 0 AN0+ Vs ground (single-ended)
  adc.setMultiplexer(ADS1115_MUX_P0_NG);
  delay(10); // adc settling delay
  current_EL = adc.getDiff0();  // Read ADC Channel 0
  if (debug_mode)
  {
    Serial.println(current_EL);
  }
  // Set the ADC to Channel 1 AN1+ Vs ground
  adc.setMultiplexer(ADS1115_MUX_P1_NG);
  delay(10); // adc settling delay
  current_AZ = adc.getDiff1();  // Read ADC Channel 1
  if (debug_mode)
  {
    Serial.println(current_AZ);
  }
}
```

The last function in our sketch is the `check_move()` function. When we receive a movement command, the command is decoded and the relays on the rotator controller are energized to move the antenna to the requested position. This function is called from the main `loop()` and continually checks the status of the rotator positioning and stops the rotation when the rotators reach the requested position.

```
// Check to see if we've been commanded to move
void check_move()
{
  // We're moving - check and stop as needed
  if (set_AZ != -1 || set_EL != -1) {
    read_adc();   // Read the ADC
    // Map AZ to degrees
    if (debug_mode)
    {
      Serial.println(current_AZ);
    }
    // Map the Current Azimuth reading to Degrees
    AZ_Degrees = map(current_AZ, AZ_0, AZ_MAX, 0, 360);
```

```
  // Map EL to degrees
  if (debug_mode)
  {
    Serial.println(current_EL);
  }
  // Map the Current Elevation to Degrees
  EL_Degrees = map(current_EL, EL_0, EL_MAX, 0, 180);
  if (debug_mode)
  {
    Serial.println(EL_Degrees);
    Serial.println(AZ_Degrees);
  }

  if (set_AZ != -1) // If Azimuth is moving
  {
    AZ_Distance = set_AZ - AZ_Degrees;  // Check how far we have to move
    // No move needed if we're within the tolerance range
    if (abs(AZ_Distance) <= AZ_Tolerance)
    {
      az_rotate_stop();  // Stop the Azimuth Rotation
      set_AZ = -1;  // Turn off the Azimuth Move Command
    } else {  // Move Azimuth - figure out which way
      if (AZ_Distance > 0)   //We need to move CW
      {
        rotate_az_cw();  // Rotate CW if positive
      } else {
        rotate_az_ccw();  // Rotate CCW if negative
      }
    }
  }
  if (set_EL != -1) // If Elevation is moving
  {
    EL_Distance = set_EL - EL_Degrees;  // Check how far we have to move
    // No move needed if we're within tolerance range
    if (abs(EL_Distance) <= EL_Tolerance) {
      el_rotate_stop();  // Stop the Elevation Rotation
      set_EL = -1;  // Turn off the Elevation Move Command
    } else {  // Move Azimuth - figure out which way
      if (EL_Distance > 0)   //We need to move CW
      {
        rotate_el_up();  // Rotate Up if positive
      } else {
        rotate_el_down();  // Rotate Down if negative
      }
    }
  }
 }
}
```

Looking back, you can see that this is a big sketch. But, by breaking it down into small blocks, it doesn't really feel that way at all. This sketch does a lot of things, but since each section is a modular block of code, you can build and test it piece by piece. Using the debug mode method, you can troubleshoot it section by section as you go. When you're finished, define the debug mode to 0, upload the sketch with the debug mode changes, and the diagnostic messages are turned off and you're ready to go.

To test the interface and set the azimuth and elevation calibration values, you can use the Arduino IDE's Serial Monitor to send the GS-232A commands directly to the Arduino. Manually rotate the azimuth and elevation rotator to zero (both meters at minimum deflection) and use the "O" and "O2" commands

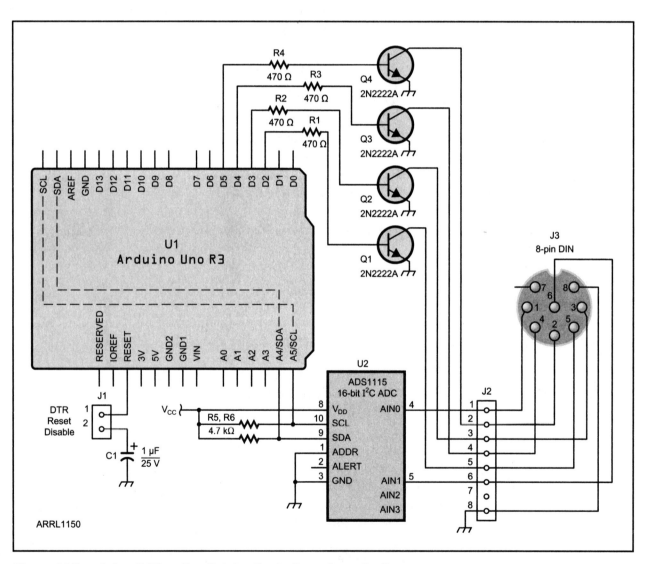

Figure 21.7 — Azimuth/Elevation Rotator Controller schematic diagram.
C1 — 1 µF, 25 V capacitor.
J1 — 2 pin header.
J2 — 8 pin header.
J3 — 8 Pin DIN female connector.
Q1-Q4 — 2N2222A NPN transistor.
R1-R6 — 470 Ω, ⅛ W resistor.
U1 — Arduino Uno.
U2 — TI ADS1115 16-bit analog-to-digital converter.
Arduino Uno Enclosure

Figure 21.8—The finished Azimuth/Elevation Rotator controller mounted in a Zigo Arduino Mega enclosure.

to set the zero calibration points. Then, manually rotate the azimuth and elevation rotators to full-scale and use the "F" and "F2" commands to set the azimuth and elevation full-scale values. Once you have done that, your controller is calibrated and the calibration values are stored in the Arduino's onboard EEPROM. You can recalibrate your controller at any time by repeating the calibration process. You can now send the various GS-232A controller commands listed back in Table 21.1 to test all the functions on your controller. When you're finished, define the debug mode to 0 and upload the sketch with the debug mode changes. This turns off the diagnostic messages and you're ready to go.

Now that your controller is calibrated, you're ready to use the controller with the *SatPC32* or *Ham Radio Deluxe* satellite tracking software on your PC. Don't forget to short the jumper on capacitor C1 to disable the DTR Reset, and have the sketch loaded with the debug mode turned off. Configure your tracking software to use the Arduino's COM port and select the Yaesu GS-232A az/el rotator controller in your tracking software. Now your G5400/5500 rotator should move and track the selected satellite.

Once you have everything working on the breadboard, you can build the circuit on an Arduino protoshield using the schematic diagram in **Figure 21.7**. Since this is a project that will be permanently located in my shack, I went for a classier look with the enclosure. I mounted my finished Azimuth/Elevation Rotator Controller in a Zigo Mega Enclosure (**Figure 21.8**) and mounted the 8-pin DIN connector on the side of the enclosure. The Zigo Arduino enclosures are available on eBay and at Amazon.com. I built an 8-pin DIN-to-DIN patch cable to interface to the Yaesu G5400/5500 rotator controller. We won't need to connect any source of external power to the Arduino since we'll be using the power provided by the PC and the Arduino's USB connector.

Enhancement Ideas

There's really not much left to do with this project in the way of enhancements. Looking at my Yaesu G5400 controller box, there may be just enough room to mount an Arduino Nano and a small circuit board inside the box. This would allow you to move the Azimuth/Elevation Rotator Controller inside the Yaesu rotator control box itself and make it completely self-contained.

Since the Yaesu GS-232 rotator computer interface is used by the Yaesu DXA-series of azimuth-only rotators, you can use this controller to drive those rotators as well. *Ham Radio Deluxe* has an option to use the Yaesu GS-232A for azimuth control only. The Yaesu DXA azimuth-only rotators use a 6-pin DIN connector instead of the 8-pin DIN connector because they don't need the two pins used for elevation in the G5400/G5500. You can even omit the elevation relay transistors used for elevation control if desired if you're going to use this project to control an azimuth-only Yaesu DXA-series rotator.

References

Amazon — **www.amazon.com**
Ebay — **www.ebay.com**
Ham Radio Deluxe — **www.ham-radio-deluxe.com**
SatPC32 software — **www.dk1tb.de**
Yaesu — **www.yaesu.com**

CHAPTER 22

CW Decoder

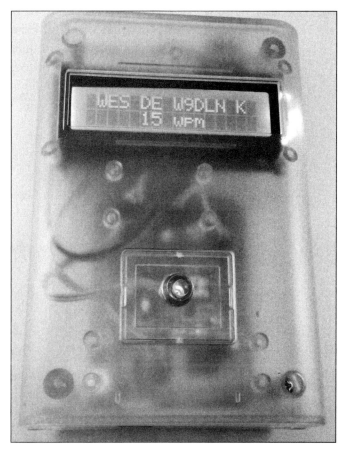

The finished CW Decoder mounted in a SparkFun pcDuino/Arduino enclosure.

How many times have you been working CW, gotten distracted and asked yourself "What did they just send?" It's times like that I wish that I had a CW decoder handy to double-check myself. Sure, I can fire up the PC with a soundcard interface and have the PC do the decoding for me, but that can be cumbersome, especially if I'm operating a portable station in the middle of nowhere without any ac power.

By now you've seen some projects that can send CW, but what about receiving and decoding CW? If they can write a computer program for the PC that uses a soundcard interface to decode CW, then surely we can do it with an Arduino. All we have to do is convert the CW signal to a digital pulse and measure the pulse widths to decode the CW. Now for the really cool part, someone has already created an Arduino library to do just that. The

`MorseEnDecoder.h` library will decode a digital CW signal into the actual letters and numbers; all we have to do is convert the CW audio tone into a digital signal that the `MorseEnDecoder.h` library can use.

Tone Decoder

To do this, we'll use a Texas Instruments LM358 op amp to buffer the CW audio, and a Texas Instruments LM567 tone decoder chip to convert the CW tone into a digital signal. By using the LM567 tone decoder, we also have the ability to limit the decoder bandwidth to help prevent interference from adjacent CW signals. The LM567 can be used to decode a tone and the output of the LM567 will go high when the desired frequency is present on the input. The center frequency and bandwidth of the decoded tone are controlled by the value of a resistor and two capacitors. To calculate the desired center frequency, you can use the formula

$$f_o = 1/(1.1 \times RC)$$

where

    R is the resistance (in ohms) across pin 5 and pin 6 on the LM567
    C is the capacitance (in farads) on the capacitor between pin 6 on the
        LM567 and ground.

The bandwidth of the tone decoder can be determined with the formula

$$BW = 1070 \times \sqrt{\frac{V_{in}}{f_o \times C}}$$

where

    V_{in} is the input signal in V_{rms}
    f_o is the center frequency
    C is the value of the capacitor (in farads) between pin 2 on LM567 and ground.

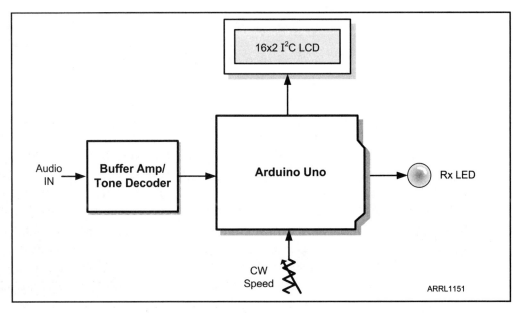

Figure 22.1 — CW Decoder block diagram.

Fortunately, there is an online LM567 tone decoder calculator at **www.vk2zay.net** to make these calculations much easier for us. Using the online calculator, a variable center frequency of 455 Hz to 909 Hz, with a bandwidth of 217 Hz to 300 Hz, was chosen for the CW Decoder project. With the potentiometer (R6) for the LM567's center frequency at a mid-range value of 5 kΩ, the center frequency will be approximately 600 Hz, with a bandwidth of 270 Hz.

Now that we know how we can convert the CW audio into a digital signal the Arduino can use, we can plan our project. The CW Decoder block diagram (**Figure 22.1**) shows that we will use the LM358/LM567 circuit to decode the CW audio signal, and we'll scroll the decoded CW on a 16-line by 2-character (16×2) LCD display. We will also use an LED on the output of the LM567 as a tuning indicator. Finally, we'll use a potentiometer to adjust the CW decode

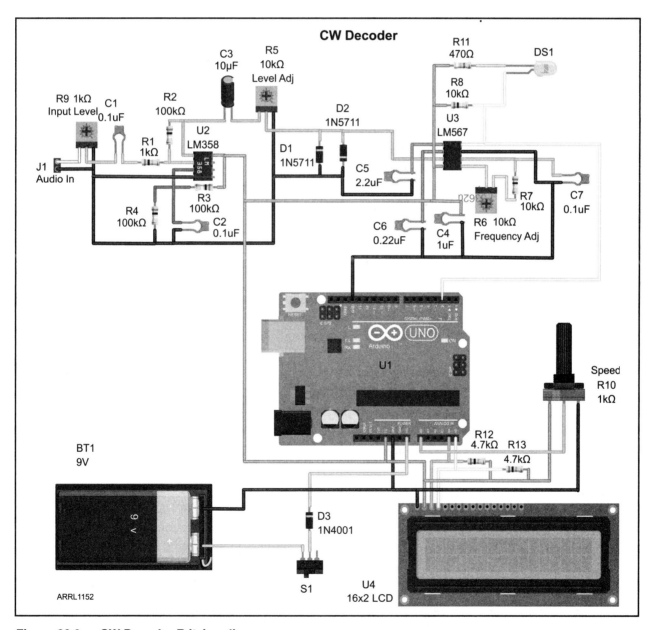

Figure 22.2 — CW Decoder Fritzing diagram.

speed range. The `MorseEnDecoder.h` library doesn't handle automatic CW decode speed adjustments, so we'll need to use the SPEED potentiometer to adjust the CW decode speed manually.

Now that we have our block diagram, we can start putting all the pieces together on a breadboard. The majority of this project involves the LM386/LM567 circuit. **Figure 22.2** shows the Fritzing diagram used to build the CW decoding circuit. Potentiometer R9 will be used to adjust the audio input to the LM358 op amp we're using to amplify and buffer the audio input to the LM567. Potentiometer R5 is used to adjust the audio input level to the LM567. The LM567 requires an input signal level of 200 mV or less to decode properly, so two 1N5711 Schottky diodes are used as clamping diodes to limit the LM567 input voltage to a maximum of 200 mV. Potentiometer R6 is used to adjust the center frequency of the CW tone to be decoded. As a starting point, adjust this resistor to its mid-range point of 5 kΩ.

Once you have the CW audio decoding circuit on the breadboard, you can test and adjust the center frequency. Connect the audio from your receiver to the audio input of the CW Decoder circuit and tune to a CW signal. You can also use the waveform generator we built back in Chapter 18 to precisely tune the center frequency of the LM567. Using an oscilloscope or voltmeter, adjust potentiometer R9 until the audio input on pin 2 of the LM358 is about 50 mV. The gain of the LM358 is controlled by the values of resistors R1 and R2. For the CW Decoder project, the gain was selected to be 100, which is calculated using the simple formula R1/R2. This will provide an output of up to 500 mV on pin 1 of the LM358. Once you have the input on pin 2 of the LM358 adjusted to 50 mV, you can adjust R5 to control the audio level into the LM567. You'll want to adjust this level to be around 200 mV on your voltmeter.

Next, tune your receiver to output the pitch of the CW tone that you prefer to listen to. Alternatively, you can use the waveform generator, and adjust its output to the frequency your ears like best, usually a frequency around 600 to 700 Hz. Adjust potentiometer R6 to find the center of the point where the LED (DS1) on pin 8 of the LM567 comes on. Your CW tone decoder circuit is now tested and tuned to the desired CW pitch.

The Sketch

Finish hooking the rest of the circuit on your breadboard and we're ready to write the sketch for the CW Decoder. Using the flowchart in **Figure 22.3**, you can see that the LM358/LM567 does most of the work for us, and the `MorseEnDecoder.h` library will handle the rest of the work on the Arduino side of things.

Starting out in the sketch, we'll define the libraries we'll need for the CW Decoder project. You'll note that the sketch defines the `avr/pgmspace.h` library, but you don't see where it is used in the sketch. The `MorseEnDecoder.h` library stores its CW conversion tables in the Arduino's flash memory, so the library itself uses the `avr/pgmspace.h` library. So, you effectively have the case of a library using a library to

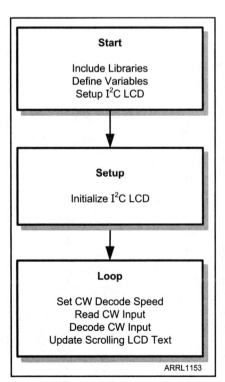

Figure 22.3 — CW Decoder flowchart.

get the job done. It doesn't get much more efficient than that. To display the decoded CW characters, we'll use a 16×2 LCD display with an I²C "backpack" to communicate with the Arduino using the I²C bus. The complete sketch can be found in **Appendix A** and online at **www.w5obm.us/Arduino**.

```
#include <avr/pgmspace.h>   // Used by the MorseEnDecoder Library
#include <MorseEnDecoder.h> // Morse EnDecoder Library
#include <Wire.h>       //I2C Library
#include <LiquidCrystal_I2C.h>  // Liquid Crystal I2C Library
```

Next, we'll define the Arduino I/O pins and the variables used in the sketch:

```
#define morseInPin 2
#define Speed_Pin A0

const int lcd_end = 16; // set width of LCD
const int lcd_address = 0x27; // I2C LCD Address
const int lcd_lines = 2; // Number of lines on LCD
String text;  // Variable to hold LCD scrolling text
char cw_rx;   // Variable for incoming Morse character
int read_speed;   // Variable for desired CW speed setting
int current_speed=-1;   // variables to track speed pot
```

Finally, we'll initialize the objects for the `MorseEnDecoder.h` and the `LiquidCrystal_I2C.h` libraries. Since we are not using the sending portion of the `MorseEnDecoder.h` library, we don't need to be concerned about the `MORSE_KEYER` or `MORSE_ACTIVE_LOW` parameters when we initialize the library.

```
// Define the Morse objects
morseDecoder morseInput(morseInPin, MORSE_KEYER, MORSE_ACTIVE_LOW);

// set the LCD I2C address to 0x27 for a 16 chars and 2 line display
LiquidCrystal_I2C lcd(lcd_address,lcd_end,lcd_lines);
```

Since the majority of the work will be handled by the `MorseEnDecoder.h` library, there's really not much to be done in the `setup()` loop. We'll initialize the LCD, turn on the backlight and display a brief startup message so that we know everything is okay to this point.

```
lcd.init();  // initialize the LCD
lcd.backlight();  // Turn on the LCD backlight
lcd.home();  // Set the cursor to line 0, column 0

lcd.print("KW5GP CW Decoder");
delay(3000);
lcd.clear();
```

In the main `loop()`, we'll first read the value of the CW receive SPEED potentiometer. If the receive speed potentiometer value has changed, the new receive speed will be displayed on the LCD and the receive decode speed will be updated in the `MorseEndecoder.h` library. Since the `MorseEnDecoder.h` library doesn't auto-detect the speed of the received CW, the SPEED potentiometer must be adjusted manually to match the receive speed. This adjustment does not have to precise. Once you start listening to the incoming CW, you can approximate the speed and adjust the receive speed accordingly. Since this is a coarse adjustment, you'll find that it's not as

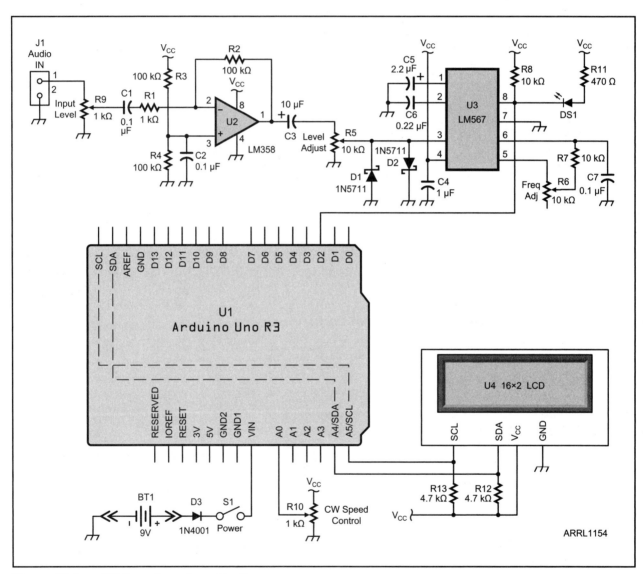

Figure 22.4 — CW Decoder schematic diagram.

BT1 — 9 V battery.
C1, C2, C7 — 0.1 µF, 16 V capacitor.
C4 — 1 µF, 16 V capacitor.
C3 — 10 µF, 16 V capacitor.
C5 — 2.2 µF, 16 V capacitor.
C6 — 0.22 µF, 16 V capacitor.
D1, D2 — 1N5711 Schottky diode.
D3 — 1N4001 diode.

DS1 — 5mm LED.
J1 — Mono mini jack.
R1 — 1 kΩ, ⅛ W resistor.
R2, R3, R4 — 100 kΩ, ⅛ W resistor.
R5, R6 — 10 kΩ trimmer potentiometer.
R7, R8 — 10 kΩ, ⅛ W resistor.
R9 — 1 kΩ trimmer potentiometer.

R10 — 1 kΩ potentiometer
R11 — 470 Ω, ⅛ W resistor.
R12, R13 — 4.7 kΩ, ⅛ W resistor.
S1 — SPST switch.
U1 — Arduino Uno.
U2 — LM358 op amp.
U3 — LM567 tone decoder.
U4 — 16×2 I²C LCD display.

cumbersome as it may seem to be at first. The `MorseEndecoder.h` library can handle some variation in the receiving speed internally, so you'll just need to set the receiving speed close to the actual speed and the library will take it from there.

```
// Read the potentiometer to determine code speed
read_speed = map(analogRead(Speed_Pin),10,1000,5,35);

// If the Speed Pot has changed, update the speed and LCD
if (current_speed != read_speed)
{
  current_speed = read_speed;  // Set the current speed to the desired speed
  morseInput.setspeed(read_speed);  // Call the set speed function
  // set up the LCD display with the current speed
  text = String(current_speed) + " wpm";
  lcd.clear();  // Clear the LCD
  lcd.setCursor(5,1);  // Set the cursor to 5,1
  lcd.print(text);  // Display the CW text
  text = "";
}
```

The actual CW decoding is done by the `MorseEnDecoder.h` library. All you have to do is call the `morseInput.decode()` function to decode the incoming character. Using the `morseInput.available()` function to see if there is a new character received, the `morseInput.read()` function will return the decoded CW character. The received characters are then scrolled from right to left on the top line of the LCD.

```
morseInput.decode();  // Decode incoming CW

if (morseInput.available())  // If there is a character available
{
  cw_rx = morseInput.read();  // Read the CW character

// Display the incoming character - Set the text to display on line 0 only.
// When length = 15, trim and add to new character so display
// appears to scroll left
if (text.length() >15)
{
  text = text.substring(1,16);  // Drop the First Character
}
text = text + cw_rx;  // Set up the text to display on the LCD
lcd.setCursor(0,0);  // Set the cursor to 0,0
lcd.print(text);  // Display the CW text
}
```

Once you have the circuit and sketch working on the breadboard, you can use the schematic diagram in **Figure 22.4** to build the finished project on an

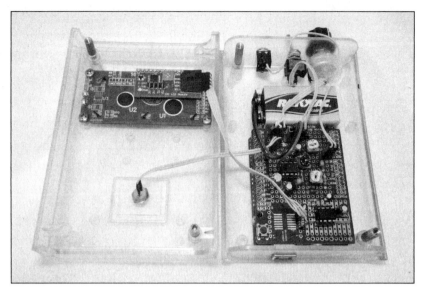

Figure 22.5 — Inside view of the CW Decoder.

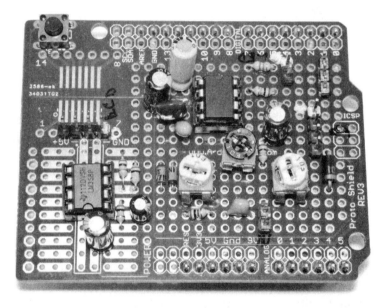

Figure 22.6 — The CW Decoder wired on an Arduino protoshield. As you can tell, this one was starting to get a little crowded.

Arduino protoshield, and mount it in an enclosure. For this project, I used a clear SparkFun pcDuino/Arduino project enclosure (**Figure 22.5**). Even though there are a lot of external components in this project, everything fits easily on the protoshield (**Figure 22.6**) with room to spare.

Enhancement Ideas

Naturally, the first enhancement that comes to mind is to downsize everything with an Arduino Nano and a small piece of perfboard, which would allow you to build the CW Decoder in a much smaller enclosure that would fit in a shirt pocket or mount to your portable QRP CW rig using hook-and-loop fasteners. You could also modify the `MorseEnDecoder.h` library, add in the code needed to implement the auto-speed detection, and have the CW decoder automatically adjust to the speed of the incoming CW signal. And, since the `MorseEnDecoder.h` library also has the ability to send CW, you could add in the code for an iambic keyer, and use the `MorseEnDecoder.h` library to turn the project into a complete CW keyer and decoder all in one small, portable package.

References

Google — **code.google.com/p/morse-endecoder/**
LM567 Tone Decoder Calculator — **www.vk2zay.net/calculators/lm567.php**
Texas Instruments — **www.ti.com**

CHAPTER 23

Lightning Detector

The finished Lightning Detector mounted in a Solarbotics Arduino Mega S.A.F.E.

Like so many of my Arduino projects, the Lightning Detector project concept started out with "Wouldn't it be cool if...?" That's one of the things I like most about the Arduino. If you can think it up, there's a good chance you can find a sensor, module, or a circuit design that will do what you want it to do. That's how it was with the Lightning Detector. Since I'm on the road a lot, I either have to remember to disconnect my antennas before I leave, or roll the dice and hope that lightning has no interest in feeding on my rigs. One evening, I was brainstorming with Tim Billingsley, KD5CKP, and I thought, "Wouldn't it be cool if you could make a lightning detector that would automatically disconnect your antennas when it detected lightning and reconnected them when the storm had passed?" The discussion went on a bit and ended with the

Figure 23.1 — The Embedded Adventures AS3935 MOD-1016 lightning detector module.

consensus that a lightning detector that did all that would be cool, but existing lightning detectors were too expensive.

That led to some research on the Internet about lightning detectors, and I stumbled across the Austriamicrosystems AS3935 Franklin lightning sensor chip. Not only that, but I found a fully assembled module that would connect directly to the Arduino. The Embedded Adventures MOD-1016 module (**Figure 23.1**) incorporates the AS3935 chip and all the support components onto a module that connects to the Arduino via either the SPI or I²C bus. The AS3935 lightning sensor can detect lightning up to 40 km away and provide distance and strength of the lightning strike. And the best part, the module only cost $21. Now that we had cleared the "too expensive" hurdle, it was time to get my hands on one of these modules and start playing.

Around the same time, I was invited by Craig Behrens, NM4T, to present a forum on the Arduino at the Huntsville Hamfest in Alabama. Craig asked me to come up with a couple of new and unique ham-related projects for the presentation. Since I had never seen a lightning detector anywhere, much less an affordable one, this would be the perfect project for the forum.

Austriamicrosystems AS3935 Franklin Lightning Detector

The Austriamicrosystems AS3935 is a programmable lightning sensor chip that can detect the presence and approach of lightning, both cloud-to-ground and cloud-to-cloud. The module can detect lightning at a distance of up to 40 km and estimates the distance to the leading edge of the storm in steps of 16 ranges. The AS3935 does this by analyzing the RF signals it receives

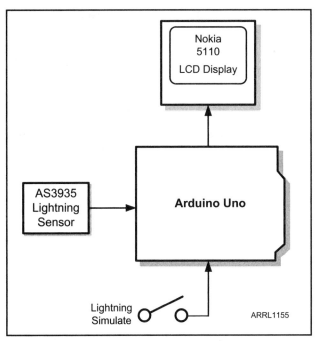

Figure 23.2 — Lightning Detector block diagram.

at approximately 500 kHz. The module has a programmable internal capacitor that is used to tune a small external loop antenna for resonance at 500 kHz. It uses a sophisticated on-chip algorithm to validate the incoming signal and can distinguish between man-made noise and lightning. If a valid lightning strike is detected, the AS3935 will calculate the energy of the strike, and then perform a statistical estimate of the distance of the strike. The module can communicate with the Arduino using either the SPI or I²C bus.

Figure 23.2 shows the block diagram for the Lightning Detector. Since you can't always get lightning to show up when you need it for testing, a LIGHTNING SIMULATE switch is added so that we can test the sketch logic and simulate a lightning strike for demonstrations.

I went back and forth over what type of display to use with this project. The small 128 × 32 pixel organic LED (OLED) display I originally used for the Huntsville Hamfest Arduino forum was really nice and allowed me to experiment with bitmap graphic images, but for demonstration purposes, the display proved to be a little on the small side. Instead, we'll use the larger Nokia display, which will allow us to show more information on the LCD and is much more readable.

When you stop and think about what this project is actually doing, it's amazing that it can be built using just nine signal wires (not including power and ground connections). Once again, this shows that you can create some pretty sophisticated projects with the Arduino without the need for a lot of external components. **Figure 23.3** shows the Fritzing diagram for the Lightning Detector project. You'll notice that we don't have any pull-up resistors on the I²C connection to the lightning sensor module. The Embedded Adventures MOD-1016 has the necessary I²C pull-up resistors

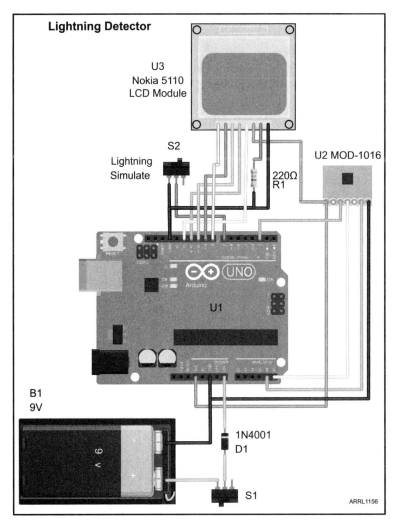

Figure 23.3 — Lightning Detector Fritzing diagram.

Lightning Detector 23-3

already installed on the module, along with a small external loop antenna the AS3935 uses to receive the 500 kHz lightning signals.

The Sketch

Writing the sketch for the Lightning Detector is almost as easy as wiring it up. The hardest part was finding an Arduino library for the I²C version of the AS3935. The Embedded Adventures MOD-1016 is pre-wired to use the I²C bus for communication. You can change the soldered jumpers on the board to use the SPI bus, but I prefer to use the I²C bus when possible, and I really didn't want to solder on the only lightning detector module I had. Most of the libraries I found online used the SPI bus to communicate with the Arduino. Once again, the Open Source community rides to the rescue. I was able to locate a modified version of the SPI-based library that works with the I²C version of the AS3935, and the sketch was good to go.

With the circuit wired up on the breadboard, we can get started on the sketch. Using the flowchart in **Figure 23.4**, you can see that the plan is to wait for a lightning event and display the information about the lightning strike on the Nokia LCD. We'll keep track of the time since the last lightning strike and display that also. We'll be using libraries for the AS3935 and the Nokia display, which will help to simply the work we have to do to create this sketch. The complete sketch for the Lightning Detector project can be found in **Appendix A** and online at **www.w5obm.us/Arduino**.

As we start out with the sketch you'll see that we use the debug mode to provide extra diagnostic information as we test and debug the sketch:

```
#define debug  1 // Enables full diagnostic reporting

// I2C and AS3935 Libraries
#include "I2C.h"   // Use the I2C Library
#include <AS3935.h>  // Use the AS4935 Library
#include <LCD5110_Basic.h>  // Use the Nokia 5110 LCD Library
#include <Wire.h>  // Include the Wire Communication Library
```

Next, we'll define the Nokia LCD and the AS3935 objects for the library.

```
/*
 It is assumed that the LCD module is connected to
 the following pins.
     CLK  - Pin 12
     DIN  - Pin 11
     DC   - Pin 10
     RST  - Pin 8
     CE   - Pin 9
*/
```

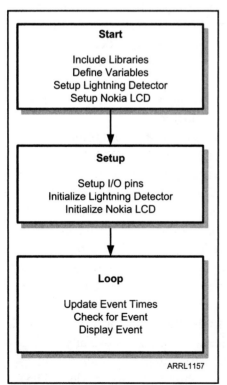

Figure 23.4 — Lightning Detector flowchart.

```
LCD5110 glcd(12,11,10,8,9); // Assign the Nokia 5110 LCD Pins

extern uint8_t SmallFont[]; // define the Nokia Font

// Lightning Detector library object initialization
// First argument is interrupt pin, second is device I2C address

AS3935 AS3935(2,3);
```

Note that we've defined the AS3935 address as 3. The AS3935 is jumper selectable to one of four different I²C address, but we'll be using the default address that the MOD-1016 module is preset to. Next, we'll define the Arduino I/O pins and AS3935 internal registers and settings we'll be using:

```
#define IRQ_PIN 2    // Define Pin 2 for the Interrupt from the Detector Module
#define SIMULATE_PIN 7  // Define Pin 7 for the Lightning Simulation switch

// Delay to allow Detector Module settling after startup
#define holddown_time 20

// Defines the Register for the High order bits of the lightning energy
#define AS3935_ENERGY_HIGH 0x6,0x1f
// Defines the Register for the Middle order bits of the lightning energy
#define AS3935_ENERGY_MID 0x5,0xff
// Defines the Register for the Low order bits of the lightning energy
#define AS3935_ENERGY_LOW 0x4,0xff

#define NoiseFloor 2   // Define the Noise Floor Level
#define SpikeReject 2  // Define the Spike Rejection Level
#define Watchdog 2     // Define the Watchdog Setting
```

In the last part of the initialization portion of our sketch, we'll define all the variables we'll need:

```
// Set the Recommended Value of the Detector Tuning Capacitor
// This is the value that comes printed on the module packaging
int recommended_tune_cap=5;

int test;  // calculated tuning cap value
int strokeIntensity;  // The intensity of the strike
int simulated;  // Indicates if strike is simulated
int irqSource;  // The source of the AS3935 interrupt
int strokeDistance;  // the distance of the strike
int holddown = 1;  // Set the Flag indicating startup
long last_event = 0;  // Holds the time of the last event detected
long last_tick =0;  // the time of the last minute "tick" in millis()
long current_time;  // the current time in millis()
// the difference in millis() between last "tick" and the last event
long time;
```

```
// Used to calcluate the intensity of the strike
long strokeEnergyHigh, strokeEnergyMid, strokeEnergyLow, strokeTotal;

// The lines of text for the Nokia display
String text1, text2, text3, text4, text5, text6 = " ";
```

In the `setup()` loop, we'll start the serial port so we can use the Arduino IDE's Serial Monitor to see the diagnostic information as we test the sketch. Then we'll start up the I²C bus and set the bus speed to the default speed of 100 kHz. We need to be sure we're at the standard I²C bus speed of 100 kHz to ensure the operating frequency of the bus is not near the 500 kHz receive frequency of the Lightning Detector. If we were to run the I²C bus in Fast Mode at 400 kHz, we run the risk of interfering with the Lightning Detector's receiver, potentially causing missed lightning events. Since we go through a tuning test and verification as part of the `setup()` loop, this portion of the sketch is a little more involved than usual.

```
Serial.begin(9600);   // set the Serial USB port speed

//I2C bus initialization
I2c.begin();
I2c.pullup(true);
I2c.setSpeed(0); //100kHz

pinMode(IRQ_PIN, INPUT);   // Setup the Detector IRQ pin
pinMode(SIMULATE_PIN, INPUT); // Setup the Lightning Simulate Button

// enable the pullup resistor on the Simulate pin
digitalWrite(SIMULATE_PIN, HIGH);

// Randomize the Arduino's random number generator
randomSeed(analogRead(0)); // seed the random number generator
simulated = LOW; // reset the simulate flag
```

Next, we'll start the Nokia 5110 LCD and display a brief startup message so we know everything is working to this point:

```
// Set up the Nokia 5110 Display
glcd.InitLCD(65);   // Initialize
glcd.setFont(SmallFont);   // Use Small Font

cleartext();   // clear the text values
text1 = "KW5GP";
text2 = "Lightning";
text3 = "Detector";
text6 = "Initializing";
updatelcd();   // update the LCD
```

In my early tests of the AS3935 communications, I had some problems verifying that the AS3935 was communicating correctly with the Arduino. The `I2C.h` library incorporates a bus scan function that will display a list of all the I²C devices on the Arduino IDE's Serial Monitor. If the debug mode is enabled, it will run the I²C bus scan function.

```
// If the debug flag is set, provide extra diagnostic information
if(debug == 1)
{
  // verify that we can see the Lightning Detector (should be at 0x3)
  Serial.println("Scan I2C Bus");
  I2c.scan();  // Run the I2c Bus Scan function
}
```

Once we have verified that the AS3935 is communicating, we'll have it perform a reset and then set up the noise floor, spike rejection and watchdog settings. We'll include a delay between commands to allow the AS3935 time to process the commands.

```
// reset all internal register values to defaults
AS3935.reset();   // Reset the AS3935
delay(1000);   // Wait a second for things to settle

// Set the Noise Floor, Spike Rejection and Watchdog settings

AS3935.setNoiseFloor(NoiseFloor);   // Set the Noise Floor Level
delay(200);
AS3935.setSpikeRejection(SpikeReject);   // Set the Spike Rejection Level
delay(200);
AS3935.setWatchdogThreshold(Watchdog);   // Set the Watchdog Level
delay(1000);
```

In the next portion of the `setup()` loop, we have some additional diagnostic code to show the results of the AS3935's tuning process. The AS3935 library includes a function to calculate the value of the capacitor used to tune the external antenna. The Embedded Adventure modules are pre-tested and the recommended value of the tuning capacitor is printed on the module package. As a general rule, you'll want to use their recommended value, but we'll include the library's tuning function as part of our diagnostic code to double check the `AS3935.h` library's tuning capacitor function's suggested value against the recommended value.

```
// The Embedded Adventures MOD-1016 includes the
// recommended Tuning Capacitor Setting on the package
// This Debug calibration routine is for testing and verification purposes

if (debug == 1)   // run calibration if debug flag set
{
```

```
  // if lightning detector cannot tune tank circuit to required tolerance,
  // calibration function will return false
  if(!AS3935.calibrate())
  {
    Serial.println("Tune Error");
  }
  delay(500);  // Wait for things to settle
  // read and display the current Tuning Cap value
  test = AS3935.registerRead(AS3935_TUN_CAP);
  delay(500);
  Serial.print("Tuning Cap: ");
  Serial.println(test,HEX);
}
```

Once we have verified the tuning capacitor value, we'll set it to the recommended value, and read it back from the AS3935 to verify that the setting is correct:

```
// Set the Tuning Cap register to the value recommended
Serial.println("Set Tune Cap Reg");
// Write the recommended value to the Tuning Capacitor Register
AS3935.registerWrite(AS3935_TUN_CAP,recommended_tune_cap);
delay(500);  // Wait for things to settle
// verify the tuning cap value is set correctly
test = AS3935.registerRead(AS3935_TUN_CAP);
delay(500);
Serial.print("Tuning Cap: ");
Serial.println(test,HEX);
```

Next, you will see a block of code that has been commented out. The AS3935 uses three internal oscillators to generate clocking signals. The LC oscillator (LCO) runs at the resonant frequency of the antenna and tuning capacitor, which should be around 500 kHz. The system RC oscillator (SRCO) typically runs at 1.1 MHz and the timer RC oscillator (TRCO) runs at 32.768 kHz. The AS3935 has the ability to send the output of these three internal oscillators to the IRQ pin for monitoring. The AS3935 library uses the output of the LCO as part of the function to calculate the tuning capacitor value. You can use an oscilloscope or a frequency counter to measure and fine-tune these settings if desired. In normal operation these oscillators are calibrated at initialization by the AS3935.h library and you'll find that the library calibration settings work just fine.

```
/*
Serial.println("Set Disp LCO Reg 8 Bit 7");
AS3935.registerWrite(AS3935_DISP_LCO,1);
delay(10000);
AS3935.registerWrite(AS3935_DISP_LCO,0);
Serial.println("LCO disp complete");
```

```
Serial.println("Set Disp SRCO Reg 8 Bit 6");
AS3935.registerWrite(AS3935_DISP_SRCO,1);
delay(10000);
AS3935.registerWrite(AS3935_DISP_SRCO,0);
Serial.println("SRCO disp complete");
*/
```

In the last portion of the `setup()` loop, we'll finish setting up the AS3935, set the receiver gain for indoor use, disable the Disturber (EMI) interrupts since we're only looking to see lightning and not man-made noise events (more on this later), and print the contents of the important AS3935 registers to the Serial Monitor one last time to ensure everything is set up correctly. We'll then clear the interrupts on the AS3935 to be sure we have a clean start with no false lightning events, and display a "No Activity" message on the Nokia LCD.

```
// tell AS3935 we are indoors, for outdoors use setOutdoors() function
AS3935.setIndoors();   // Set Gain for Indoors
// AS3935.SetOutdoors;  // uncomment to set Gain for Outdoors
delay(200);   // Wait for things to Settle

// Uncomment to turn on Disturber Interrupts (EMI)
//AS3935.enableDisturbers();
// We only want Lightning, turn off EMI interrupts
AS3935.disableDisturbers();
delay(200);   // Wait for it things to settle
printAS3935Registers();   // Display the AS3935 registers

// clear the IRQ before we start
int irqSource = AS3935.interruptSource();
delay(500);

Serial.println("Detector Active");  // And we're ready for lightning

cleartext();   // Clear all the LCD text variables
text3 = "No Activity";
updatelcd();   // Update the LCD
```

In the main `loop()`, we'll start out with a one minute timing check. The Lightning Detector will update the Nokia LCD once a minute and indicate the time since the last event occurred. If a lightning strike is detected, its information is immediately displayed and the time since the last event is restarted.

```
// Check and update timestamp
current_time = abs(millis())/1000;   // Current time (seconds since start)
if (current_time - last_tick >= 60)  // Run if 60 seconds has passed
{
  // One Minute has passed
```

```
  Serial.print("Tick  ");
  Serial.print(current_time);  // Print the current time
  Serial.print("   ");
  Serial.print(last_tick);  // Print the previous time
  Serial.print("   ");
  last_tick = current_time;
  // convert to minutes and hours
  // convert difference last event to current time into seconds
  time = last_tick - last_event;
  Serial.print(time);

  if (time >=60 ) // One minute has passed
  {
    time = time/60;
    text6 = String(time);
    if ((time >= 1) && (time <60))
    // 1 to 59 minutes ago
    {
      text6 = text6 + " min ago";
    } else {
      if (time >=60)
      {
        time = time/60;
        // convert to hours
        text6 = String(time);

        if (time >=1)
        {
          if (time == 1)
          {
            text6 = text6 + " Hour ago";
          } else {
            text6 = text6 + " Hours ago";
          }
        }
      }
    }
    updatelcd();  // Update the LCD with time since last event
  }
  Serial.println("    " + text6);
}
```

Before we allow the Lightning Detector to start detecting events, we add a brief "hold down" time immediately after starting up to allow the AS3935 to settle out and build its initial statistics that it uses in the lightning calculations.

```
// Delay the start for a few seconds to let everything settle
if (holddown == 1)
{
  if ((millis()/1000) > holddown_time)  // If we've passed the hold down time
  {
    holddown = 0;  // Turn it loose
  }
} else {     // Hold Down has already expired, watch for lightning
```

Now we're ready and waiting for a lightning or simulation event. First, we'll check and see if the LIGHTNING SIMULATE switch has been pressed:

```
// Check for a simulation
if (digitalRead(SIMULATE_PIN) == LOW)
{
  simulated = HIGH;  // Turn on the simulated flag
  // disable the simulate button for 1 second to debounce the switch
  delay(1000);
  Serial.print("Sim Button ");
} else {
  simulated = LOW;  // Make sure we turn off the simulated flag
}
```

Next, we'll check to see if the AS3935 has generated any interrupts:

```
// If we have a real or simulated event let's do it
if ((digitalRead(IRQ_PIN) == HIGH) || (simulated))
{
  // if it's a real event, use the actual interrupt
  // otherwise set the IRQ code for lightning
  if (!simulated) // It's not simulated, get the IRQ from the AS3935
  {
    delay(200);  // wait for interrupt register to settle
    irqSource = AS3935.interruptSource();  // Read the AS3935 IRQ Register
    Serial.print("Real Irq ");
  } else {
    // It's a simulation - pretend it's actual lightning
    irqSource = 0b1000;  // Set the IRQ code for lighting
  }
  Serial.print("IRQ: ");
  Serial.print(irqSource);
  Serial.print(" ");
```

Now we'll process the interrupt type. The AS3935 has four types of interrupts. An interrupt value of 0 (no IRQ register bits set) indicates that the AS3935 has purged its internal statistics. This is a normal event that occurs about 15 minutes after the last interrupt event occurs. An interrupt value of 1 (IRQ register bit 0 is set) indicates that a Noise High interrupt has occurred, while an interrupt value of 4 (IRQ register bit 2 is set) indicates that a Disturber

(EMI — man-made noise) interrupt has occurred. An interrupt value of 8 (IRQ register bit 3 is set) indicates that lightning has been detected. If you have the Disturbers disabled, the only interrupts the AS3935 will generate are the Statistics Purge and Lightning Detected events.

When an AS3935 interrupt is detected, we create a timestamp so we can track how long ago the event occurred.

```
// The first step is to find out what caused interrupt
// as soon as we read interrupt cause register, irq pin goes low
// The returned value is bitmap field, bit 0 - noise level too high,
// bit 2 - disturber detected, and finally bit 3 - lightning!

// Create timestamp so we know when it occurred
// IRQ value of 0 is a stat purge, we don't want to do anything with it
if(irqSource != 0)
{
  // Create a timestamp for the event
  timestamp();  // Run the timestamp function
  text6= "Now";  // Set the Text time of the event to "Now"
}
```

Now we'll start decoding the interrupts and displaying the results to the Nokia LCD. The first two interrupts are the Noise High and Disturber (EMI) interrupts. We don't do anything with this other than display the type of interrupt and how long ago the last event occurred:

```
if (irqSource & 0b0001)   // Noise Level High Interrupt
{
  text3 = "Noise High";
}

if (irqSource & 0b0100)   // Man Made Disturber (EMI) Interrupt
{
  text3 = "EMI Detect";
}
```

Here's where the fun starts. We'll handle the interrupt for lightning. This can either be an actual lightning interrupt from the AS3935, or it can be a simulated lightning event. The AS3935 will report the estimated distance of the strike in 16 ranges.

```
if (irqSource & 0b1000)   // It's Lightning or a Simulation
{
  // We need to find distance of the lightning strike,
  // The function returns the approximate distance in kilometers,
  // where value 1 represents storm in detector's near victinity,
  // and 63 - very distant, out of range stroke
  // everything in between is the distance in kilometers
```

23-12   Chapter 23

```
  text2 = "Detected";
  if (simulated) // make up a distance if we're faking it
  {
    strokeDistance = int(random(45));  // Pick a distance between 1 and 44)
    // If a real strike is less than 5km, the AS3935 will report that
    // it's "Overhead", so we match that
    if (strokeDistance < 5 )
    {
      strokeDistance = 1;
    }
    text1 = "Simulation";
  } else {    // otherwise, it's lightning get the real distance
    // It's real lightning, read the AS3935 Distance Register
    strokeDistance = AS3935.lightningDistanceKm();
    text1 = "Lightning"
  }
```

Once we get the distance, we'll format this information for display at the next Nokia LCD update:

```
// The AS3935 Reports Lightning distance as Out of Range,
// 40,37,34,31,27,24,20,17,14,12,10,8,6,5,Overhead
if (strokeDistance < 5)
{
  text3 = "Overhead";
}
if (strokeDistance > 40)
{
  text3 = "Out of Range";
}
if (strokeDistance <= 40 && strokeDistance >= 5)
{
  text3 = String(strokeDistance) + " km away";
}
```

Next, we'll get the energy of the detected strike. If this is a simulation, we'll make up the numbers. Once we have calculated the estimated energy of the strike, we'll scale the energy value from 1 to 10 to create an arbitrary "Lightning Intensity Factor." We'll then format this information to display the next time we update the Nokia LCD.

```
if (simulated)   // Make up the energy of the stroke
{
  // There are 3 registers containing strike energy
  strokeEnergyHigh = int(random(31));
  strokeEnergyMid = int(random(255));
  strokeEnergyLow = int(random(255));
} else {    // otherwise get the real energy
```

```
    // Read the 3 AS3935 Strike Energy Registers
    strokeEnergyHigh = AS3935.registerRead(AS3935_ENERGY_HIGH);
    strokeEnergyMid = AS3935.registerRead(AS3935_ENERGY_MID);
    strokeEnergyLow = AS3935.registerRead(AS3935_ENERGY_LOW);
}
// Calculate the total Strike Energy
strokeTotal = strokeEnergyLow + (strokeEnergyMid*256)+ (strokeEnergyHigh*65536);
// Map the energy to an arbitrary Intensity Factor of 1 thru 10
strokeIntensity = map(strokeTotal,1,2000000,1,10);
text4 = "Intensity: " + String(strokeIntensity);
```

Finally, we'll output the information to the Nokia LCD.

```
if (irqSource == 0)   // It's a Statistics Purge, we ignore these
   {
   // The AS3935 will Purge old lightning data automatically
   Serial.print(" Stats Purged ");
} else {
   updatelcd();   // Update the LCD with the Event Data
   Serial.print("Time: ");
   Serial.print(" ");
   Serial.print(last_event);
   Serial.print(" "+text1);
   //   Serial.print(text3);

   // Display the information if it's a lightning event
   if (irqSource & 0b1000)
   {
     Serial.print("  Energy: ");
     Serial.print(String(strokeTotal));
     Serial.print("  Dist: ");
     Serial.print(strokeDistance);
     Serial.print(" ");
     Serial.print("Intensity: ");
     Serial.print(strokeIntensity);
     Serial.print(" ");
}
```

There are four functions used to support the Lightning Detector sketch. The printAS3935Registers() function is used to display the contents of the key AS3935 registers. The timestamp() function is used to create a timestamp to keep track of the time of an event. The updatelcd() will clear the Nokia LCD and display the current contents of the LCD text variables. The cleartext() function is used to clear the contents of the LCD text variables in preparation for the next display update.

```
void printAS3935Registers()      // Display the basic AS3935 Registers
{
```

```
  // Read the Noise Floor setting
  int noiseFloor = AS3935.getNoiseFloor();

  // Read the Spike Rejection setting
  int spikeRejection = AS3935.getSpikeRejection();

  // Read the Watchdog Threshold setting
  int watchdogThreshold = AS3935.getWatchdogThreshold();

  Serial.print("Noise floor: ");
  Serial.println(noiseFloor,DEC);
  Serial.print("Spike reject: ");
  Serial.println(spikeRejection,DEC);
  Serial.print("WD threshold: ");
  Serial.println(watchdogThreshold,DEC);
}

// stores the millis() time of the last event in seconds
void timestamp()
{
  last_event = abs(millis()/1000);
}

// clears LCD display and writes the current LCD text data
void updatelcd()
{
  glcd.clrScr();
  glcd.print(text1,CENTER,0);
  glcd.print(text2,CENTER,8);
  glcd.print(text3,CENTER,16);
  glcd.print(text4,CENTER,24);
  glcd.print(text5,CENTER,32);
  glcd.print(text6,CENTER,40);
}

// clears the LCD text data
void cleartext()
{
  text1 = " ";
  text2 = text1;
  text3 = text1;
  text4 = text1;
  text3 = text1;
  text4 = text1;
  text5 = text1;
  text6 = text1;
}
```

Now, here's the hardest part of the whole project — finding lightning to detect. When I finished building my Lightning Detector prototype on the breadboard, I spent weeks waiting for a lightning storm to pass through. I used the LIGHTNING SIMULATE button to test things, but it's just not the same without real lightning. Finally the day came when I woke up to lightning, and up popped all the lightning events. It was a bit difficult to debug things and figure out what an interrupt value of zero (Statistics Purge) was, and why I kept getting it 15 minutes after lightning was last detected. It's documented in the AS3935 datasheet, buried as one line in the fine print. Once I had that figured out, it was time to move everything into a nice enclosure for demonstrations.

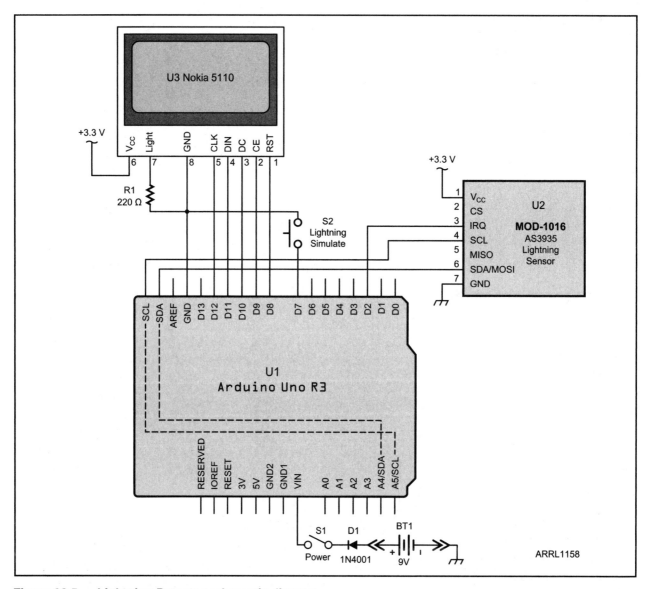

Figure 23.5 — Lightning Detector schematic diagram.

D1 — 1N4001 diode.
R1 — 220 Ω, ⅛ W resistor.
S1 — SPST switch.
S2 — SPST momentary contact switch.
U1 — Arduino Uno.

U2 — Embedded Adventures MOD-1016 lightning sensor.
U3 — Nokia 5110 LCD display.
Arduino Uno Enclosure

Careful Packaging is Needed

Using the schematic diagram in **Figure 23.5**, the finished project was built on an Arduino protoshield, mounted on top of the Arduino Uno, and everything was mounted in a Solarbotics Arduino Mega S.A.F.E. enclosure. Then it was back to waiting for the next thunderstorm to test the finished project. Unfortunately, it was weeks before the next storm, and the discovery of the last little glitch in the project.

I had originally placed everything inside the enclosure, mainly to protect the Nokia LCD and the fragile-looking antenna on the AS3935 module. Let's back up and review how the AS3935 actually works. For all intents, it is a narrowband receiver tuned for 500 kHz. In my initial design, I had mounted the AS3935 module inside the enclosure near the Arduino Uno. As it turns out, the Uno generates your typical low-grade RF noise from its 16 MHz oscillator and various clock signals, including the I²C bus. The Nokia 5110 LCD display also generates some RF noise as it scans and displays the pixels on the LCD. These factors combined to interfere with the reception on the AS3935 module, and it would not detect lightning very well at all. The performance was nowhere near as good as it had been when it was on the breadboard. I tried various types of displays and configurations, having to wait for thunderstorms after each new design was mounted in the enclosure.

As a last resort, I moved the AS3935 outside the Arduino enclosure onto a small wooden mast to provide some added distance from the noisemakers inside the enclosure. At one point, I had three different versions of the Lightning Detector sitting around the house waiting for a storm. As luck would have it, a fairly rare winter thunderstorm came rolling through one morning, and I was finally able to verify my suspicions about the RF noise and determine which design worked best. The winning design can be seen in **Figure 23.6**. As a final touch, we'll use a ping-pong or plastic golf ball to provide some measure of protection and make the project look like it has a little radome to add to the cuteness factor.

Enhancement Ideas

Now that we have a working Lightning Detector, there are all sorts of interesting features we can add. After building this project, I discovered the Parallax/Grand Idea Studios Emic 2 text-to-speech module we used in previous projects in this book. Adding voice output to the Lightning Detector project opens up a world of possibilities. You can drive a PA system at Field

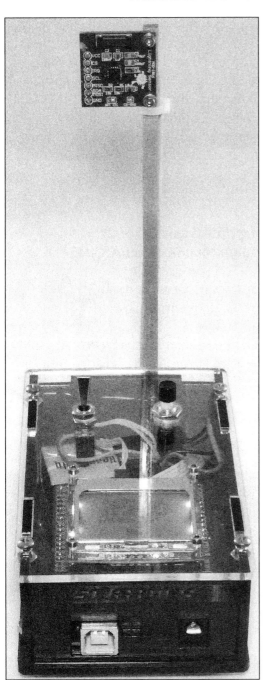

Figure 23.6 — The finished Lightning Detector module mounted on its mast.

Day to announce lightning in the area, or you can just have it on your desk to tell you when lightning is nearby. If you didn't want to use a voice module, you could have the Lightning Detector sound an alert tone when lightning is detected.

Also, I've never been all that happy with the way I handle the timing of the events and the one-minute "tick." This project would be a perfect candidate to play with the Arduino's timer interrupts, and move the timing operation into a software timer interrupt-driven function.

Since my original design concept for the Lightning Detector project called for it to disconnect my antennas in advance of a thunderstorm and reconnect them after the storm had passed, adding the ability to control an antenna relay box would complete the project for me. I didn't have a whole lot of luck finding a relay made exclusively for switching antenna coaxial cable at an affordable price, but I did find an article in the April 2005 issue of *QST* that will fit the bill nicely. "A Low-Cost Remote Antenna Switch" by Bill Smith, KO4NR (see References) looks like it will do everything my original concept called for. Once I get that put together and hooked up, I can head out of town secure in the knowledge that the Lightning Detector back at home is safeguarding my equipment.

References

Arduino I²C Master Library — **www.github.com/rambo/I2C**
AS3935 Arduino Library — **www.github.com/SloMusti/AS3935-Arduino-Library**
Austriamicrosystems AG — **www.austriamicrosystems.com**
Embedded Adventures — **www.embeddedadventures.com**
Solarbotics — **www.solarbotics.com**
B. Smith, KO4NR, "A Low-Cost Remote Antenna Switch," *QST*, Apr 2005, pp 38-41. This project also appears in the Station Accessories chapter of recent editions of *The ARRL Handbook*.

CHAPTER 24

CDE/Hy-Gain Rotator Controller

Homebuilt CDE/Hy-Gain rotator controller.

How many times have you come across an antenna rotator at a hamfest that didn't have the control box? Or maybe lightning zapped your rotator control box and now you have a working rotator motor, but no way to turn it. Replacing an antenna rotator control box can get expensive, assuming you can even find a working one.

Not long ago, a friend of mine, Shawn Braddock, W5SMB, was faced with this very dilemma. Shawn had acquired a tower with a rotator, but the rotator controller had long since gone missing. Since I've done a lot of tinkering with the Arduino and the Yaesu rotator controllers, I offered to build an Arduino-based controller for his CDE/Hy-Gain HAM-III rotator. This was going to be a fun challenge. I would have to build the controller completely from scratch based on the original HAM-III schematics I managed to find online, and adapt the rotator for Arduino control. (HAM series rotators have been used by amateurs for decades. They were originally made by Cornell Dubilier Electronics — CDE — and are now made by Hy-Gain, which is now part of MFJ. The HAM-M, HAM-II, HAM-III, HAM-IV, HAM-V and Tailtwister models all will work with this Rotator Controller project.)

I really didn't want to just duplicate the existing controller with an analog position meter and some switches. If there's an Arduino under the hood, why not use it and kick things up a notch? The plan is to use an analog-to-digital

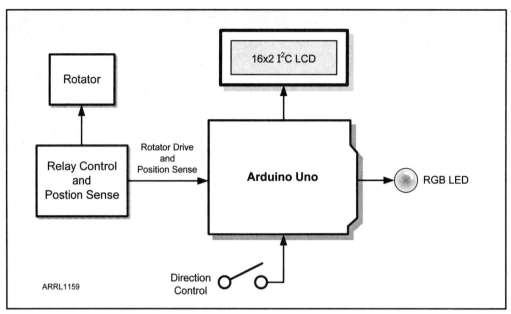

Figure 24.1 — Rotator Controller block diagram.

(A/D) converter to sense the rotator position, and have the Arduino drive relays to control the rotator brake and activate the drive motors. Because we're controlling all this with an Arduino, let's add in a brake delay that's not in the standard CDE/Hy-Gain rotator controller. We'll replace the analog meter with a 16-character by 2-line (16×2) LCD display, plus we'll add an RGB LED to show the drive motor and brake status.

Now that we know what we're planning to have our rotator controller do, we'll start out with the block diagram in **Figure 24.1**. We'll use a two-position center-off switch to tell the Arduino which way to turn the rotator. To do the actual motor and brake control, we'll have the Arduino drive three 5 V relays with 10 A contacts. For the rotator position sensing, we'll use a Texas Instruments ADS1115 4-channel 16-bit I²C A/D converter to read the voltage coming in from the 500 Ω variable resistor mounted inside the rotator housing. Finally, we'll display the antenna heading on a 16×2 LCD in place of the analog meter used in the CDE/Hy-Gain control box. We'll have the Arduino display the antenna bearing in degrees and for fun, we'll also have it display direction arrows when the rotator is turning, and display a message when the rotator is braking. We'll also use an RGB LED to indicate the drive relay status just for kicks.

Packaging

Since we're going to need an enclosure for the main rotator power transformer and all the rest of the goodies, we can't use a standard Arduino enclosure like we've been using. Putting all this in an Altoids mint tin is certainly out of the question. For this project, we're going to need a bigger box. The HAM-III rotator brake solenoid and motors run on 26 V ac at about 3 A. I bought a used HAM-III controller power transformer online from C.A.T.S. for $25 since there wasn't much hope of finding one locally. That turned out to be the most expensive part of the whole project. That's part of the fun about the

Arduino. You can create some very complex projects and they're not going to cost you an arm and a leg. The total cost for this entire project ended up around $80. That's less than the cost of a used CDE/Hy-Gain controller, and you won't have anywhere near the fun and sense of accomplishment. Not to mention, the standard CDE/Hy-Gain controllers don't have a brake delay or PC interface, so computer control isn't even a possibility without spending a whole lot more money. Once we have the basic controller up and running, all it takes is a little extra software to add in the interface for *Ham Radio Deluxe* or other rotator control software for your PC.

The next order of business was to find an enclosure large enough to house the transformer and everything else mounted inside. All of the enclosures available locally were just a little too short to house the power transformer. After searching around for an enclosure big enough to hold everything, but not so big as to put a dent in my budget, I found the perfect selection of enclosures on TEN-TEC's website. As it turns out, not only does TEN-TEC make radios, they also sell a complete line of very affordable enclosures. For this project, I chose the TP-49 enclosure in the aluminum finish at a cost of $12. The TP-49 measures 3.25 × 7.75 × 6.25 inches (height, width, depth), perfect for what we've got in mind.

Now that we've got all the major pieces figured out, it's time to start building our rotator controller. This will be a two-pronged approach. We'll need to mount everything inside the enclosure, but, since the enclosure is a plain box, we're going to need to cut some holes and figure out how to mount things. Before we can start to build the rotator controller, we'll need to design the circuit and wire everything inside the enclosure. Because we're dealing with high voltage, we're not going to test this project on the breadboard. Instead, we'll build everything directly in the enclosure, and do all our testing with the actual finished controller.

Circuit Details

Even though we're not going to use the breadboard for this project, I still like to create both a schematic diagram and a Fritzing diagram. Since a Fritzing diagram is more of a physical representation of the circuit, I find it easier to use the Fritzing diagram as a wiring guide and check things off on it as I go. I like to use the schematic diagram in the final steps of construction to verify all the connections and do any circuit troubleshooting that may be needed.

Figure 24.2 shows the Fritzing diagram for the CDE/Hy-Gain Rotator Controller project. Don't let the apparent complexity fool you. Many of the components will be mounted on an Arduino protoshield that will be plugged onto the Arduino expansion shield headers, and we'll use header pins and sockets for wiring that connects to the external relays, LCD, switch, and other parts. Since we want this to be a standalone rotator controller, we'll also add in a small 12 V power supply to provide the current needed to power the Arduino and drive the relays. The circuit may look daunting, but you'll find that the actual wiring of the project goes smoothly and easily. The hardest part is cutting all the holes and mounting everything in the enclosure.

Since we're going to be wiring everything in the enclosure, we'll also want to have the schematic diagram (**Figure 24.3**) handy. Before we can begin the actual wiring, we'll need to figure out how and where we're going to mount

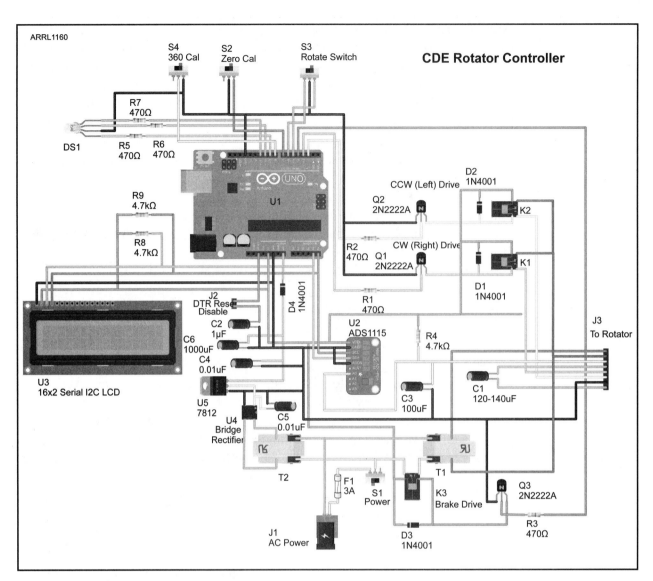

Figure 24.2 — Rotator Controller Fritzing diagram.

Figure 24.3 — Rotator Controller schematic diagram.
C1 — 120-140 µF, 220-250 V ac motor capacitor.
C2 — 1 µF, 16 V capacitor.
C3 — 100 µF, 16 V capacitor.
C4, C5 — 0.01 µF, 35 V capacitor.
C6 — 1000 µF, 35 V capacitor.
D1-D4 — 1N4001 diode.
DS1 — 5 mm RGB common cathode LED.
F1 — 3 A fuse.
J1 — AC power plug.
J2 — 2 pin jumper header.
J3 — 8 pin header to connect to rear panel connector to rotator.
K1-K3 — 5 V SPST relay, 10 A contacts.
Q1-Q3 — 2N2222A transistor.
R4, R8, R9 — 4.7 kΩ, ⅛ W resistor.

R1, R2, R3, R5, R6, R7 — 470 Ω, ⅛ W resistor.
S1 — SPST switch.
S2, S4 — SPST momentary pushbutton switch.
S3 — SPDT switch, center off.
T1 — CDE/Hy-Gain 26 V ac rotator controller transformer.
T2 — 12.6 V transformer.
U1 — Arduino Uno.
U2 — ADS1115 16-bit A/D converter module.
U3 — 16×2 I²C LCD.
U4 — Bridge rectifier.
U5 — LM7812 voltage regulator.
Fuse Holder
Ten-Tec Model TP-49 Enclosure

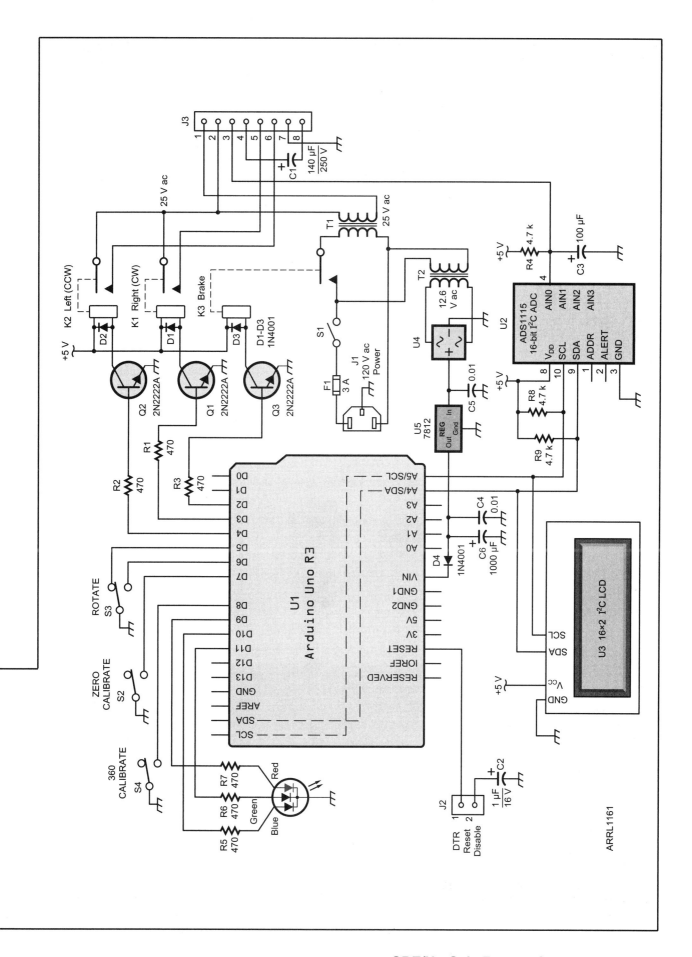

everything in the enclosure. Since this involves cutting holes, we'll start by preparing the enclosure.

First, we'll cut out a hole for the 16×2 LCD and find a place to mount the big power transformer. I used a drill and a nibbler to cut out the holes to the approximate shape and then filed the area smooth. Fortunately, the aluminum enclosure was easy to work with and the hole-cutting went smoothly. **Figure 24.4** shows the results.

Figure 24.4 — Rotator Controller chassis showing power transformer mounting and LCD cutout.

Figure 24.5 — Rotator Controller chassis rear view.

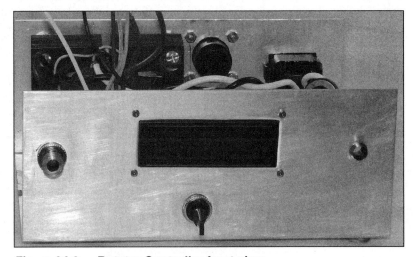

Figure 24.6 — Rotator Controller front view.

On the rear of the enclosure (**Figure 24.5**), we'll need to cut out the holes for the ac power socket, fuse holder, rotator connector, and the Arduino's USB connector. We'll want the Arduino's USB connector brought out to the back of the box so we can hook up a PC to upload the sketch and for interfacing with *Ham Radio Deluxe* or other rotator control software at a later date.

Now that we have the locations of the big parts laid out, the next step is to mount the LCD, POWER switch, ROTATE switch, and RGB LED holder to the front of the chassis. Four 2 mm socket head screws were used to mount the LCD in its front panel cutout. For the ROTATE switch, a large double-pole double throw center-off momentary switch was mounted below the LCD assembly. **Figure 24.6** shows the finished front panel.

Figure 24.7 — Pigtail cable used between rotator controller and the CDE/Hy-Gain rotator.

The large power transformer is mounted on the rear panel flange of the chassis itself, along with the ac power socket, fuse holder, and the AMP/TYCO CPC circular socket for the rotator connector. The rotator connector will connect to a male circular connector on a short pigtail cable that has Anderson Powerpole connectors to connect to the rest of rotator cable (**Figure 24.7**). Since I use Powerpoles to standardize all my rotator connections, the pigtail allows for quick and easy swapping between the Arduino project and a standard CDE/Hy-Gain rotator controller during test and setup.

Figure 24.8 shows all of the external components mounted inside the chassis with the Arduino Uno mounted to the bottom of the chassis with standoffs. The four standoffs

Figure 24.8 — Rotator Controller chassis showing the Arduino mounting point.

in front of the Arduino will be used to mount the perfboard containing the 12 V transformer and voltage regulator we'll be using to generate the supply voltage used to power the Arduino. Since the Rotator Controller is designed to operate without a PC connected to the Arduino's USB port, we'll be using this power supply to power the Arduino portion.

The three relays used to drive the brake and rotator motors are epoxy glued to the top of a small piece of perfboard and mounted on the inside front edge of the controller box underneath the power switch (**Figure 24.9**).

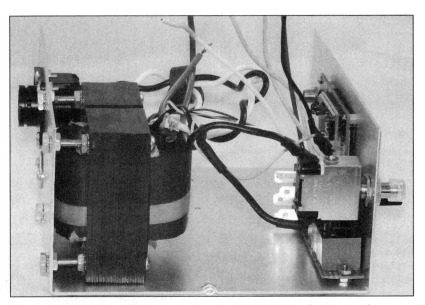

Figure 24.9 — The relay board mounted behind the front panel underneath the power switch.

Figure 24.10 — Rotator Controller chassis with all internal components mounted.

The 12 V power transformer and voltage regulator are installed on a square piece of perfboard and mounted on top of the standoffs in front of the Arduino (**Figure 24.10**). The ZERO and 360° CALIBRATION SET pushbutton switches are also mounted on the low voltage power supply board. This board will connect to the Arduino using the right-angle header pins located on the lower right side of the board.

Now that all the major components have been mounted, it's time to build the Arduino protoshield that will be used to interface between the all of the

Figure 24.11 — The assembled Rotator Controller protoshield.

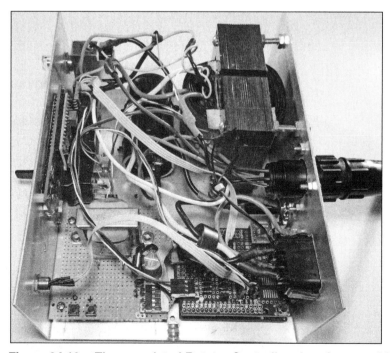

Figure 24.12 — The completed Rotator Controller chassis assembly.

external components mounted in the chassis and the Arduino. The protoshield will be used to mount the relay drive transistors, the ADS1115 A/D module, and the other components needed to interface to the rest of the controller. The components mounted in the chassis connect to the protoshield using DuPont-style header and socket pin connectors to allow for easy installation and removal of the protoshield. The finished protoshield is shown in in **Figure 24.11**.

Figure 24.12 shows the completed wiring of the Rotator Controller chassis. DuPont-style header sockets have been attached to the leads coming from the external components to the prototyping shield and the low voltage power supply is connected. The wiring for the rotator controller is now complete. You can see that while the wiring looks complex, it isn't really all that hard when you take it in small steps. For me, the hardest part was figuring out where to mount everything and to get the holes cut properly.

Software Design

Just as with the chassis construction, the sketch for this project looks complex, but when you break it down into small functional chunks, you'll see that it's really not that difficult. In fact, some of the sketch came directly from the Azimuth/Elevation Rotator Controller project and other sketches we've created to this point. This is yet another aspect that helps make developing projects fun. If you build your sketches in a modular fashion, often you can re-use that code in another sketch, saving you a lot of time by not having to reinvent the wheel.

Figure 24.13 shows the flowchart we'll use to create the CDE/Hy-Gain Rotator Controller sketch. When you put aside the thoughts of all that complex wiring you just did, you'll see that we don't need the Arduino to do all that much actual work. Once we have everything initialized, the Arduino will read the rotator position and wait for the ROTATE switch to be pressed. Once the ROTATE switch has been pressed, the Arduino will energize the rotator brake release and the appropriate drive motor relay. As the rotator is turning, the current heading is updated on the LCD, along with an arrow to indicate the direction of rotation.

When the ROTATE switch is released, the motor relay is de-energized, but the brake solenoid is held energized for an additional three seconds to allow the rotator and antenna to coast to a stop. The brake solenoid is then de-energized and the brake locks the rotator into position.

While the Arduino is waiting for the ROTATE switch to be pressed, it will also check to see if one of the CALIBRATION SET pushbutton switches has been pressed. These switches are used to save the A/D converter's values for the current rotator position in the Arduino's EEPROM for the 0° and 360° rotator position.

Reading the Rotator Position

On my prototype version, everything was going smoothly right to the point where I tried to read the rotator position while the rotator was turning. When everything was idle, the position sensing was absolutely perfect. The problems started when I tried to read the rotator position potentiometer when

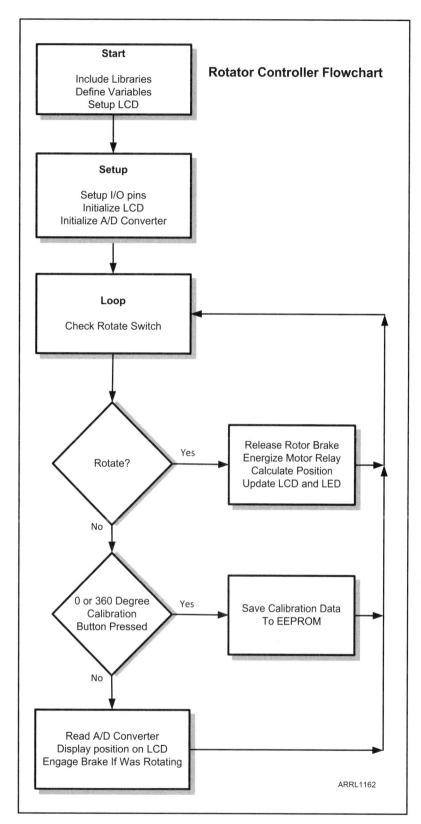

Figure 24.13 — Rotator Controller sketch flowchart.

the rotator was turning. The HAM series rotators use 26 V ac to drive the rotator motors and brake solenoid while the rotator is operating. The 26 V ac used for the brake and motors shares its ground with the 12 V dc used for the rotator position potentiometer mounted inside the bell housing. With the analog meter in the standard CDE/Hy-Gain rotator controllers, this was not a concern. However, when you try to read the position potentiometer with a high-speed, high-resolution A/D converter, there is about a half of a volt worth of ac hum on the ground. I tried every trick I knew to eliminate the ac hum from the dc positioning sensor with no success. After all the work I had done to this point, it was not looking good for this project at all. In fact, I ended up putting it off to the side to revisit later, and continued on with the other work.

While researching this project, I had been able to find only one other Arduino-based CDE/Hy-Gain rotator controller project, and that one required opening up the rotator bell housing and separating the position potentiometer dc ground from the motor ac ground. While that resolved the ac hum issue, I did not want a project that required you to go up on your tower and modify your rotator assembly. These are supposed to be relatively easy projects, and my design criteria called for no major modifications, if any, to the rotator assembly itself.

A potential answer came out of the blue in an on-air discussion that I had with Tim Billingsley, KD5CKP. Unknowingly, Tim planted the seed for the solution while we were talking about how the standard CDE/Hy-Gain controller for the HAM-series rotators wasn't all that accurate, and he would generally move the antennas to position, wait for things to settle, and then move to the final position. Looking at the standard rotator controller with an oscilloscope and voltmeter, I saw that it also had the ac hum on the position potentiometer, but it was only affecting the analog position meter on the regular controller by a couple of degrees. It turns out the analog meter was seeing the ac hum, but since it was 60 Hz hum, you couldn't really see it on the front panel meter of the rotator control box.

So, this meant that reading the position with the A/D converter while the rotator was turning was pretty much out of the question. It looked like that was the end of the road for this project unless I wanted to modify the position sensor potentiometer connections inside the rotator housing. Then, Tim's comment that he moved his antennas to a coarse antenna bearing, and then did a second move to the final position finally fired enough brain cells to turn on the old light bulb.

Let's stop and think about the basics of how the rotator does what it does. At the fundamental level, all that is happening is that the motors are turned by a gear assembly at a constant rate. Using a stopwatch, I timed the 0 to 360° rotation for the HAM-III rotator on my workbench, and found that it turned the full 360° in an average of 49 seconds. Doing some quick math, that comes out to about 7.347° of rotation per second.

I had been wanting to experiment with the Arduino software timer interrupts, but had not really found a suitable project for them, until now. By using a software timer interrupt, I could basically do the old dead-reckoning time-versus-rotation distance calculation trick, and estimate the rotator position while the rotator was turning, instead of trying to read the A/D values of the position potentiometer. After some quick testing to get the feel for how the Arduino software timer interrupts worked, I incorporated the time and rotational distance calculations into the interrupt handler function in the sketch and began my tests. Once I got a few minor bugs ironed out in the sketch, my first run had the rotator controller tracking the real position within 3° of the actual position over a 360° rotation. Fine tuning brought that down to less than 1° of difference between the calculated position and the actual position of the rotator. Further testing determined that a 250 ms interrupt was the optimum interrupt time to update the front panel LCD display and accurately calculate the rotator position while it was turning. So, now we had a design that worked and met all the design criteria.

The Sketch

Now it's time to create the final sketch we're going to use to finish up the Rotator Controller. The complete sketch and libraries used in this project can be found in **Appendix A** or online at **www.w5obm.us/Arduino**. Starting out in the sketch, we'll include all the libraries we'll need. Note that the libraries for the ADS1115 A/D converter and the I2Cdev library that supports the ADS1115 did not correctly handle the 16-bit data from the A/D and had to be modified. When you create your sketch, be sure to download the modified version of these libraries from **www.w5obm.us/Arduino**.

```
// Enable Debug Mode - Send Debug output to Serial Monitor
#define debug_mode 1

#include <Wire.h>     //I2C Library
#include <LiquidCrystal_I2C.h>  // Liquid Crystal I2C Library
#include <EEPROM.h>   // Include EEPROM Library

// Include ADS1115 ADC Library - Library Updated to fix errors
#include "ADS1115.h"

// Include I2Cdev Library - Library Updated to fix errors
#include "I2Cdev.h"

#include "Timer.h"  // Timer Library so we can do timed interrupts
```

Next, we'll define the I²C LCD. We'll also define the speed for the Serial Monitor so that we can use it to help debug the sketch and watch what's going on while the rotator is turning.

```
#define lcd_end 16 // set width of LCD
#define lcd_address 0x27 // I2C LCD Address
#define lcd_lines 2 // Number of lines on LCD

const int comm_speed = 9600;   // Set the Serial Monitor Baud Rate
Next, we'll define all the Arduino I/O pins that we'll need to control the relays
and the RGB LED inside the controller chassis:
```

```
#define left 2      // Assign Left (Counter Clockwise) Relay to Pin 3
#define right 3     // Assign Right (Clockwise) Relay to Pin 2
#define brake 4     // Assign Brake Relay to Pin 4
#define rotate_CW 5    // Assign CCW switch to Pin 5
#define rotate_CCW 6   // Assign CW switch to Pin 6
#define cal_zero 7     // Assign Zero Calibrate switch to Pin 7
#define cal_360 8      // Assign 360 Calibrate Switch to Pin 8
#define red 9     // Assign the Red LED to Pin 9
#define blue 10   // Assign the Blue LED to Pin 10
#define green 11  // Assign the Green LED to Pin 11
```

In order to get the A/D converter to read the rotator position potentiometer accurately, a 4.7 kΩ pull-up resistor is attached to one side of the potentiometer that comes in on pin 3 of the rotator connector. We want to feed a low current into the position potentiometer so we don't cause it to burn out when it's at the lower resistance positions. Due to the way the current in the pull-up/position potentiometer circuit will vary as the resistance of the position potentiometer changes, we'll need a pull-up resistance value of approximately 10 times the highest resistance of the 500 Ω potentiometer to prevent any distortion of the position voltage coming in from the position potentiometer. This results in a position-sensing voltage that will vary from 0 to approximately 400 mV depending on the rotator position.

This is another reason why I prefer to use the ADS1115 A/D module instead of the Arduino's onboard A/D converter. The ADS1115 has six programmable gain settings. For this project, we'll set the gain for a maximum of 0.512 V (512 mV).

```
#define adc_gain 0x04    // Set ADC Gain to 0 to 0.512 volts
```

Next, we'll define the brake delay. We'll allow three seconds after the motors have been turned off for the rotator and antenna to coast to a stop before we engage the brake solenoid to lock the rotator in place.

```
#define brake_delay 3000   // Set the Brake Delay to 3 seconds
```

Now we'll define the data format we'll use to store the calibration values in the Arduino's EEPROM and define the default A/D values if there is no calibration data saved:

```
#define EEPROM_ID_BYTE 1   // EEPROM ID to validate EEPROM data location
#define EEPROM_ID 55   // EEPROM ID Value
#define EEPROM_AZ_CAL_0 2    // Azimuth Zero Calibration EEPROM location
#define EEPROM_AZ_CAL_MAX 4  // Azimuth Max Calibration Data EEPROM location

#define AZ_CAL_0_DEFAULT 0
#define AZ_CAL_MAX_DEFAULT 27000
```

The next step is to define all the variables needed in our sketch. You will note that there are separate rotation speed values for when the rotator is turning clockwise (right) or counterclockwise (left). On the HAM-III rotator used in my testing, there was enough variation in the rotational speeds that I felt the need for different values for each direction to maintain position accuracy while the rotator was turning.

```
int current_AZ;  // Variable for current Azimuth ADC Value
int AZ_Degrees;  // Variable for Current Azimuth in Degrees
boolean moving = false;  // Variable to let us know if the rotor is moving

// Variable to let us know that rotation timing has started
boolean timing = false;

int set_AZ;  // Azimuth set value
int AZ_0;  // Azimuth Zero Value from EEPROM
int AZ_MAX;  // Azimuth Max Value from EEPROM

int tickEvent;
// Integer variable for position calculation while moving
int derived_degrees;

float calculated_degrees;  // Variable for position calculation while moving

// Left rotational speed in degrees per 250ms
float left_rotate_speed = 1.8341;

// Right rotational speed in degrees per 250ms
float right_rotate_speed = 1.8369;

String direction;
```

In the last part of the initialization portion of our sketch, we'll initialize the objects for the LCD, the ADS1115 and the software timer interrupt.

```
// set the LCD I2C address to 0x27 for a 16 chars and 2 line display
LiquidCrystal_I2C lcd(lcd_address,lcd_end,lcd_lines);

Timer t;  // Create Timer object as t

ADS1115 adc;  // Define as adc
```

Starting out in the `setup()` loop, we'll start the Serial Monitor port and the LCD. We'll then display a brief startup message to let us know everything is good to this point.

```
Serial.begin(comm_speed);  // Start the Serial port

lcd.init();  // initialize the LCD
lcd.backlight();  // Turn on the LCD backlight
```

```
lcd.home();   // Set the cursor to line 0, column 0

lcd.print("Rotor Controller");   // Display startup message
lcd.setCursor(4,1);
lcd.print("by KW5GP");
```

Next, we'll set all the Arduino I/O pin modes and make sure that everything is turned off before we start.

```
// Define the Input and Output pins
pinMode(right, OUTPUT);   // Set the Right Control Relay pin for Output
pinMode(left, OUTPUT);   // Set the Left Control Relay pin for Output
pinMode(brake, OUTPUT);   // Set the Brake Solenoid Relay pin for Output
pinMode(rotate_CCW, INPUT);   // Set the Rotate CW Switch pin for Input
pinMode(rotate_CW, INPUT);   // Set the Rotate CCW Switch pin for Input
pinMode(cal_zero, INPUT);   // Set the Zero Degree Calibrate Switch for Input
pinMode(cal_360, INPUT);   // Set the 360 Degree Calibrate Switch for Input
digitalWrite(rotate_CCW, HIGH);   // Enable the internal pull-up resistors
digitalWrite(rotate_CW, HIGH);
digitalWrite(cal_zero, HIGH);
digitalWrite(cal_360, HIGH);
pinMode(red, OUTPUT);   // Set the RGB LED pins for Output
pinMode(blue, OUTPUT);
pinMode(green, OUTPUT);
ledOff();   // Turn off the RGB LED
```

Now we'll read the Arduino's EEPROM to load the 0° and 360° calibration data. If the debug mode is enabled, we'll output the calibration values to the Serial Monitor.

```
read_eeprom_cal_data();   // Read the EEPROM calibration data

if (debug_mode)   // Display the calibration data when in debug mode
{
  Serial.print("Calbration Values - Zero = ");Serial.print(AZ_0);
  Serial.print(" Max = "); Serial.println(AZ_MAX);
}
```

Finally, in the last part of the `setup()` loop, we'll configure and start the ADS1115:

```
// Set the ADC sample speed to 128 samples/sec
adc.setRate(ADS1115_RATE_128);
delay(100);   // wait for the ADC to process command

adc.setGain(adc_gain);   // Set the ADC gain
delay(100);   // wait for the ADC to process command

// Set the ADC Channel 0 to single ended mode
```

```
adc.setMultiplexer(ADS1115_MUX_P0_NG);
delay(100);  // adc settling delay
```

Starting out in the main `loop()` of the sketch, we'll update the software timer interrupt so it can handle the next interrupt if the rotator is moving:

```
t.update();   // Update the Interrupt handler
```

If the rotator is not moving, we'll read the current rotator position and display it on the LCD:

```
// Read the ADC normally as long as the motor or brake are not energized
if (!moving)
{
  // Read ADC and display position when we're not moving
  // Value is placed in current_AZ
  read_adc();

  // Map the current_AZ to calibrated degrees
  AZ_Degrees = map(current_AZ, AZ_0, AZ_MAX, 0, 360);

  // Send position to Serial Monitor in debug mode
  if (debug_mode) {Serial.println(AZ_Degrees);}

  lcd.setCursor(6,0);   // Display the current position on the LCD
  if (AZ_Degrees <0)    // Set position to 0 degrees if below 0
  {
    AZ_Degrees = 0;
  }
  if (AZ_Degrees > 360)  // Set position to 360 degrees if above 360
  {
    AZ_Degrees = 360;
  }
  if (AZ_Degrees < 100)  // Adjust spacing on LCD
  {
    lcd.print(" ");
  }
  lcd.print(AZ_Degrees);  // Display the position on the LCD
  lcd.print(char(223));   // Add the degree symbol
  lcd.print("   ");
```

If the rotator is turning, we check to see if it had previously been moving or if this is the start of a move. If this is the start of a move, we'll set up the software timer interrupts for 250 ms, and allow the interrupt handler to do the rotational distance calculations while the rotator is turning.

```
} else {
  // We're moving, ADC reading is useless - too much noise
  if (!timing)   // Check to see if we're already timing the event
```

```
    {
      // We're just starting to move - start the interrupt timer
      tickEvent = t.every(250, Tick);  // Set to interrupt every 250ms
      if (debug_mode)   // Send Interrupt start message in debug mode
      {
        Serial.print("250ms second tick started id=");
        Serial.println(tickEvent);
      }
      timing = true;   // Indicates we're timing the move
      calculated_degrees = AZ_Degrees;   // Set the starting point for the move
    }
  }
```

Since the software timer interrupt can happen at any time in the loop and can take care of itself, we can move on in the `loop()` and check the rotate switch to see if it has been released. If the rotator is not turning, we'll read the rotate switch to see if it has been pressed and determine which direction we need to turn the rotator.

```
// Read the rotate switch
// Check to see if the Rotate Switch is activated
if (digitalRead(rotate_CCW) == LOW || digitalRead(rotate_CW) == LOW)
{
  // Check for Move Right (CW) Switch
  if (digitalRead(rotate_CW) == LOW && !moving)
  {
    // Rotate CW
    move_right(); // Call the Move Right function
  }
  // Check for Move Left (CCW) Switch
  if (digitalRead(rotate_CCW) == LOW && !moving)
  {
    // Rotate CCW
    move_left();  // Call the Move Left function
  }
} else {
  if (moving)   // If we were moving, time to stop
  {
    moving = false;   // Turn off the moving flag
    ledRed();   // Turn the LED Red to indicate Braking Cycle
    all_stop();   // Call the Stop Rotation function
  }
}
```

In the last portion of the main `loop()`, we'll check to see if either of the CALIBRATION SET pushbuttons have been pressed. If a CALIBRATION SET button has been pressed, we'll save the A/D value for the current position in the Arduino's EEPROM.

```
// Read the Zero Degree Calibrate Switch
if (digitalRead(cal_zero) == LOW)
{
  //  Cal Zero switch pressed
  // Set the current position as the zero calibration point
  AZ_0 = current_AZ;
  write_eeprom_cal_data();  // Write the Calibration data to the EEPROM

  if (debug_mode)  // Display Calibration Data in debug mode
  {
    Serial.print("Zero Azimuth Calibration Complete -  Zero = ");
    Serial.print(AZ_0); Serial.print(" Max = "); Serial.println(AZ_MAX);
  }
}

// Read the 360 Degree Calibrate Switch
if (digitalRead(cal_360) == LOW)
{
  //  Cal 360 switch pressed
  AZ_MAX = current_AZ -25;  // Adjust top end down a bit to allow for jitter
  write_eeprom_cal_data();  // Write the Calibration Data to the EEPROM
  if (debug_mode)  // Display the Calibration Data when in debug mode
  {
    Serial.println("Max Azimuth Calibration Complete - Zero = ");
    Serial.print(AZ_0); Serial.print(" Max = "); Serial.println(AZ_MAX);
  }
}
```

There are 12 functions used in this sketch. By moving the actual control activities into functions, the sketch can be assembled in a building block fashion. This allows you to debug your Rotator Controller one piece at a time. It also gives you a functional building block that you can use in future sketches, saving you the time of having to rewrite the same block of code over and over again.

The `all_stop()` function is used to turn off the motor drive relays, wait three seconds, then engage the rotator brake. A message indicating we are in the brake delay cycle will be displayed on the front panel LCD and the RGB LED will be turned red during the braking cycle.

```
// Relay Off function - stops motion
// Delays 3 seconds then turns off Brake Relay
void all_stop()
{
  lcd.setCursor(0,1);  // Display Braking message on LCD
  lcd.print("    Braking     ");
  digitalWrite(right, LOW);  // Turn off CW Relay
  digitalWrite(left, LOW);   // Turn off CCW Relay
  timing = false;  // turn off the timing flag
```

```
  direction = "S";      // Set direction to S (Stop)
  t.stop(tickEvent);    // Turn off the Timer interrupts
  delay(brake_delay);   // Wait for rotor to stop
  digitalWrite(brake, LOW);  // Engage the Rotor Brake
  lcd.setCursor(0,1);   // Clear the Braking Message
  lcd.print("                ");
  ledOff();    // Turn off the LED
}
```

The `move_left()` and `move_right()` functions activate the brake solenoid and the motor drive relays to turn the rotator. The `moving` flag is turned on and an arrow indicating the direction of rotation will be displayed on the LCD. If we're turning left, the RGB LED will glow green; if we're turning right it will glow blue.

```
void move_left()   // Turn the rotor Left (CCW)
{
  ledGreen();   // Turn on the Green LED
  moving = true;   // set the moving flag
  direction = "L";   // set the direction flag to "L" (Left)
  lcd.setCursor(0,1);   // display the left arrow on the LCD
  lcd.print(char(127));
  digitalWrite(brake, HIGH);   // Release the Brake
  digitalWrite(left, HIGH);    // Energize the Left Drive Relay
}

void move_right()   // Turn the rotor Right (CW)
{
  ledBlue();   // Turn on the Blue LED
  moving = true;   // set the moving flag
  direction = "R";   // set the direction flag to "R" (Right)
  lcd.setCursor(15,1);   // display the right arrow on the LCD
  lcd.print(char(126));
  digitalWrite(brake, HIGH);   // Release the Brake
  digitalWrite(right, HIGH);   // Energize the Right Drive Relay
}
```

The next block of functions controls the RGB LED. Since there are three digital I/O pins controlling the RGB LED, it makes sense to move the LED controls into functions, rather than have to repeat the sequence of three `digitalWrites()` inside the sketch to set the appropriate LED color or turn the LED off.

```
// LED Off function
void ledOff()
{
```

```
  digitalWrite(red, LOW);   // Set all RBG LED pins High (Off)
  digitalWrite(green, LOW);
  digitalWrite(blue, LOW);
}

// RED LED ON function
void ledRed()
{
  digitalWrite(red, HIGH);   // Turn on the RGB Red LED On
  digitalWrite(green, LOW);
  digitalWrite(blue, LOW);
}

// Green LED ON function
void ledGreen()
{
  digitalWrite(red, LOW);
  digitalWrite(green, HIGH);   // Turn on the RGB Green LED On
  digitalWrite(blue, LOW);
}

// Blue LED ON function
void ledBlue()
{
  digitalWrite(red, LOW);
  digitalWrite(green, LOW);
  digitalWrite(blue, HIGH);   // Turn on the RGB Blue LED On
}

// White LED function
void ledWhite()   // Turn on all LEDs to get white
{
  digitalWrite(red, HIGH);
  digitalWrite(green, HIGH);
  digitalWrite(blue, HIGH);
}
```

Next, we have the `read_adc()` function to read the 16-bit A/D converter. If debug mode is enabled, the value of the current rotator position is sent to the Serial Monitor.

```
// Read the A/D Converter
void read_adc()
{
 // display ADC read status in debug mode
 if (debug_mode) {Serial.print("Read ADC Function  ");}

  // Set the ADC sample rate to 128 samples/second
```

```
  adc.setRate(ADS1115_RATE_128);
  delay(10);   // Wait for ADC to settle

  adc.setGain(adc_gain);   // Set the ADC gain
  delay(10);   // Wait for ADC to settle

  adc.setMultiplexer(ADS1115_MUX_P0_NG);   // Set the ADC to single-ended mode
  delay(100);   // Wait for ADC to settle and start sampling

  current_AZ = adc.getDiff0();   // Read ADC channel 0

  // Display ADC value in debug mode
  if (debug_mode) {Serial.println(current_AZ);}
}
```

The `read_eeprom_cal_data()` function is used to read the saved position calibration data from the Arduino's EEPROM into the sketch variables. If debug mode is enabled, the calibration values will be sent to the Serial Monitor.

```
void read_eeprom_cal_data()   // Read the EEPROM Calibration data
{
  // Verify the EEPROM has valid data
  if (EEPROM.read(EEPROM_ID_BYTE) == EEPROM_ID)
  {
    if (debug_mode) // Display the Calibration data in debug mode
      {
        Serial.println("Read EEPROM Calibration Data Valid ID");
        Serial.println((EEPROM.read(EEPROM_AZ_CAL_0) * 256) + EEPROM.read(EEPROM_AZ_CAL_0 + 1),DEC);
        Serial.println((EEPROM.read(EEPROM_AZ_CAL_MAX) * 256) + EEPROM.read(EEPROM_AZ_CAL_MAX + 1),DEC);
      }
    // Set the Zero degree Calibration Point
    AZ_0=(EEPROM.read(EEPROM_AZ_CAL_0)*256)+EEPROM.read(EEPROM_AZ_CAL_0 + 1);
    // Set the 360 degree Calibration Point
    AZ_MAX = (EEPROM.read(EEPROM_AZ_CAL_MAX)*256) + EEPROM.read(EEPROM_AZ_CAL_MAX + 1);

  } else {
  // EEPROM has no Calibration data - initialize eeprom to default values
  if (debug_mode)
  {
    // Send status message in debug mode
    Serial.println("Read EEPROM Calibration Data Invalid ID - setting to defaults");
  }
  AZ_0 = AZ_CAL_0_DEFAULT;   // Set the Calibration data to default values
```

```
    AZ_MAX = AZ_CAL_MAX_DEFAULT;
    write_eeprom_cal_data();  // Write the data to the EEPROM
  }
}
```

The `write_eeprom_cal_data()` function is used to write the current position calibration data to the Arduino's EEPROM and updates the sketch variables with the new calibration data.

```
// Write the Calibration data to the EEPROM
void write_eeprom_cal_data()
{
    if (debug_mode)
  {
    // Display status in debug mode
    Serial.println("Writing EEPROM Calibration Data");
  }
  // Write the EEPROM ID to the EEPROM
  EEPROM.write(EEPROM_ID_BYTE,EEPROM_ID);
  // Write Zero Calibration Data High Order Byte
  EEPROM.write(EEPROM_AZ_CAL_0,highByte(AZ_0));
  // Write Zero Calibration Data Low Order Byte
  EEPROM.write(EEPROM_AZ_CAL_0 + 1,lowByte(AZ_0));
  // Write 360 Calibration Data High Order Byte
  EEPROM.write(EEPROM_AZ_CAL_MAX,highByte(AZ_MAX));
  // Write 360 Calibration Data Low Order Byte
  EEPROM.write(EEPROM_AZ_CAL_MAX + 1,lowByte(AZ_MAX));
}
```

Our last function is the interrupt handler function that is called when a software timer interrupt occurs. Whenever a timer interrupt occurs, the Arduino pauses execution of the main `loop()` and executes this function. When the function is completed, the Arduino resumes the main `loop()` right where it left off before it received the interrupt.

```
// Timer Interrupt Handler
void Tick()
{
  if (debug_mode)  // Display Interrupt information in debug mode
  {
    Serial.print("250ms second tick: millis()=");
    Serial.print(millis());
    Serial.print("    ");
  }
  // If Interrupts are enabled, it means that we're in the process of
  // timing the rotation - Add the estimated distance traveled
  // to the current calculated positon
```

```
    if (direction == "R") // If we're moving Clockwise
    {
      // Increase when we move right
      calculated_degrees = calculated_degrees + right_rotate_speed;
    } else {
      // Decrease when we move left
      calculated_degrees = calculated_degrees - left_rotate_speed;
    }
    if (debug_mode)  // Display Calculated Rotation information in debug mode
    {
      Serial.print("    ");
      Serial.print("Rotating ");
      Serial.print(direction);
      Serial.print("   ");
      Serial.println(calculated_degrees);
    }
    lcd.setCursor(6,0);   // Update the LCD with the estimated position
    derived_degrees = (int)calculated_degrees;
    if (derived_degrees < 0)  // Set to Zero if we calculate below zero
    {
      derived_degrees = 0;
    }
    if (derived_degrees > 360)  // Set to 360 if we calculate above 360
    {
      derived_degrees = 360;
    }
    if (derived_degrees < 100)
      {
        lcd.print(" ");
      }
    lcd.print(derived_degrees);  // Display the position on the LCD
    lcd.print(char(223));   // Add the degree symbol
    lcd.print("   ");
}
```

Since every rotator controller turns at a slightly different speed, you'll need to tune the left and right rotate speed variables to get your rotator controller to accurately display the current position while moving. You will need to do this with the debug mode disabled, as the slight delay introduced by sending data to the Serial Monitor can affect the timing for the position calculations.

To tune the right rotation speed variable, I turned the rotator from 0° to approximately 340° and compared the final estimated value to the actual position displayed when the brake cycle completed. I used 340° in case we were estimating too high, and our sketch modifies any values higher than 360 to be 360°. I then repeated the process from 360° to 20° to adjust the left rotation speed variable. Be sure to let the rotator cool after a few full rotations. It's not made for constant duty and may overheat if you turn it too much while you're tuning the rotation speed variables.

Enhancement Ideas

The 0°/360° position on our rotator controller starts at the far left (counterclockwise) rotation point. The standard CDE/Hy-Gain Rotator Controller has the 0°/360° position at mid-scale on the meter. If you prefer the 0/360° point to be at the midpoint of the rotation, you can modify the sketch to mirror the values of the standard ontroller by using the same method we used in the Azimuth/Elevation Rotator Controller project to modify the position values.

The sketch for this project does not include the Yaesu GS-232A azimuth-only rotator emulation code to allow your PC to automatically control the rotator. You can add that functionality if you want to use *Ham Radio Deluxe* or other antenna positioning software on your PC to automatically control the rotator. The Reset Disable jumper needed to interface with *Ham Radio Deluxe* has been included in the design for this project, in case you want to add that code in. To upload your sketch, the Reset Disable jumper must be removed, otherwise the Arduino IDE can't automatically reset the Arduino using the DTR signal to enter the sketch upload mode. Since *Ham Radio Deluxe* toggles the DTR signal line when it initially attempts to connect to the rotator, it inadvertently resets the Arduino and the rotator controller will never connect to *Ham Radio Deluxe*. If you are adding the interface for *Ham Radio Deluxe*, remember to add the Reset Disable jumper after uploading your sketch.

References

Rotor Parts by C.A.T.S — **www.rotor-parts.com**
TEN-TEC — **www.tentec.com**

CHAPTER 25

Modified CDE/Hy-Gain Rotator Controller

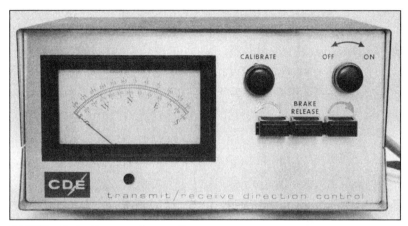

CDE/Hy-Gain Rotator Controller.

The CDE/Hy-Gain HAM series is probably the most popular antenna rotator ever made. However, computer control for the CDE/Hy-Gain rotators didn't come about until the HAM-V series in 1994. This doesn't help those of us who have the older versions. By now you may have guessed that I have a thing for the Arduino and rotator controllers, so it would be a natural to modify an existing CDE/Hy-Gain controller and give it Arduino-powered computer control by *Ham Radio Deluxe* (*HRD*) and other rotator control programs.

In researching this project, I found only one similar Arduino project for the CDE/Hy-Gain controllers, but that project required modifications inside the rotator bell housing assembly. This is due to the basic design of the rotator assembly as discussed in the previous rotator controller project in Chapter 24. Because the positioning sensor inside the rotator bell housing shares its dc ground with the rotator brake solenoid and drive motor ac ground, there is about a half a volt of ac hum on the positioning signal that I have been unable to filter out. That is why we ended up doing the time versus rotational distance calculations to estimate the rotator position in the previous project.

For this project, I took a slightly different approach. After playing with my CDE/Hy-Gain controller, measuring voltages and looking at signals all through the rotator positioning circuit, I found that the analog position meter filtered out a large percentage of the ac hum. In fact, when measuring the positioning voltage across the analog meter, the signal was nearly usable. With the addition of a filtering capacitor across the analog meter, the ac hum on the positioning signal was almost totally gone and the analog-to-digital (A/D) converter was able to reliably read the rotator position signal, even with the rotator turning. Adding the capacitor across the meter did not affect the analog meter readings at all, and may actually have made it more accurate as well.

With every bit of good news, there comes a little bad. While the position sensing signal was now usable from a stability standpoint, because we're measuring across the meter assembly, the voltage is at a very low level — in the range of 0 to 55 mV. Once again, the Texas Instruments ADS1115 A/D converter rides to the rescue. The ADS1115 has the ability to read two channels of signals differentially or four channels in single-ended mode. By using the differential mode, we can have the ADS1115 read the positioning signal across the analog meter in the rotator control box. The ADS1115 also has six programmable gain settings. As luck would have it, the lowest gain setting on the ADS 1115 is a full scale reading of 256 mV. This means that at the maximum rotator position indication of 55 mV, the A/D converter would read about $\frac{1}{5}$ of its full scale reading, or approximately 7040 A/D counts. This yields an overall resolution of about 19.5 counts per degree, which should be sufficient for positioning the rotator accurately.

Since we want this project to be able to interface with the rotator controller in *Ham Radio Deluxe*, we'll have the Arduino emulate the Yaesu GS-232A in azimuth-only mode. Since we already created a sketch that emulates the Yaesu GS-232A when we did the Azimuth/Elevation Rotator Controller in an earlier chapter, we'll be able to re-use the majority of that sketch here, reducing the sketch development process drastically. In fact, so much of that sketch's code was able to be re-used that the initial test sketch for this project was up and running in just a few minutes. This is yet another reason the Arduino has become so popular. Because the sketch for the Azimuth/Elevation Rotator Controller project was created using groups of function calls, with just a few minor changes, those same functions could be adapted for an azimuth-only controller such as the CDE/Hy-Gain unit.

Modifying the Stock Control Box

Figure 25.1 shows the block diagram for the Modified CDE/Hy-Gain Rotator Controller project. The actual modification to the rotator control box is minimal, consisting of a small board to hold the three control relays and an Arduino Uno with a protoshield to interface to the relays and position sensor. We'll use the Arduino's USB port to interface with the PC running *Ham Radio Deluxe*.

Starting out, we have to find a place inside the CDE/Hy-Gain rotator controller box to mount the Arduino Uno. The only place with enough room is on the underside of the controller box, between the 26 V ac power transformer and the motor capacitor (**Figure 25.2**). We'll cut out a small hole in the back cover to allow access to the Arduino's USB port for programming and interfacing to the PC. The downside to mounting the Arduino in the only

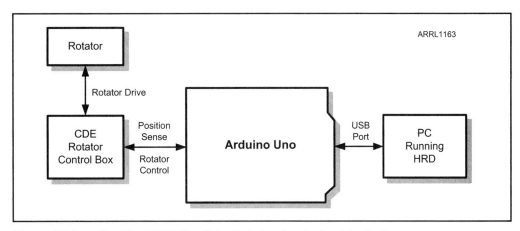

Figure 25.1 — Modified CDE/Hy-Gain Rotator Controller block diagram.

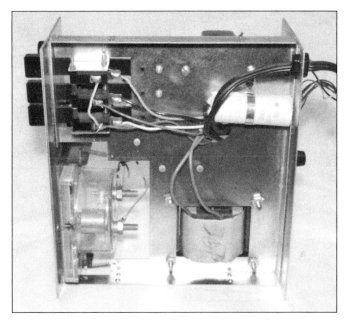

Figure 25.2 — The underside of the CDE/Hy-Gain rotator controller. The Arduino will be mounted in the open space between the power transformer and the motor capacitor.

Figure 25.3 — Rear view of the modified CDE/Hy-Gain rotator controller showing the cutout for the Arduino Uno's USB port.

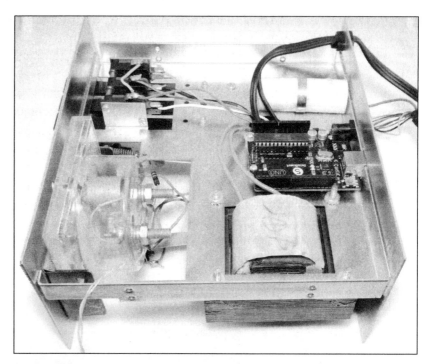

Figure 25.4 — The Arduino Uno mounted in the CDE/Hy-Gain rotator controller.

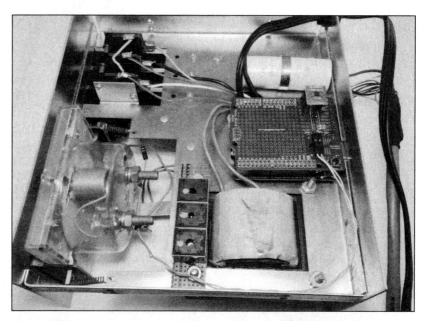

Figure 25.5 — The brake and motor relays mounted in the chassis.

available space is that the cutout for the USB port is also where the control box chassis has a double layer of metal, which makes the hole cutting just a slightly more difficult (**Figure 25.3**). Fortunately it's only a small hole we need to cut out, so it goes fairly quickly. **Figure 25.4** shows the Arduino Uno mounted to the underside of the control box chassis.

Next, we'll use epoxy to glue the three miniature relays to a strip of perfboard and mount it to the ends of two existing screws in the control box chassis (**Figure 25.5**). You can mount the relay board anywhere you want on the underside of the chassis, but the two existing screws were just too convenient to ignore.

Now it's time to wire everything together. Because we need to wire in the switches and position sensing circuit inside the rotator control box, we won't build a test circuit on the breadboard. Instead, we'll wire everything in place in the rotator control box and hope for the best. Be sure and unplug your control box, as there is 120 V ac all over the inside. **Figure 25.6** shows the Fritzing diagram for modifying your CDE/Hy-Gain rotator controller. As with the Yaesu Azimuth/Elevation Rotator Controller, only a couple of components are needed to build this project. We'll mount the resistors and transistors used to drive the relays, along with the ADS1115 A/D converter, and the Reset Disable jumper on an Arduino protoshield. We'll then use the DuPont-style header and socket

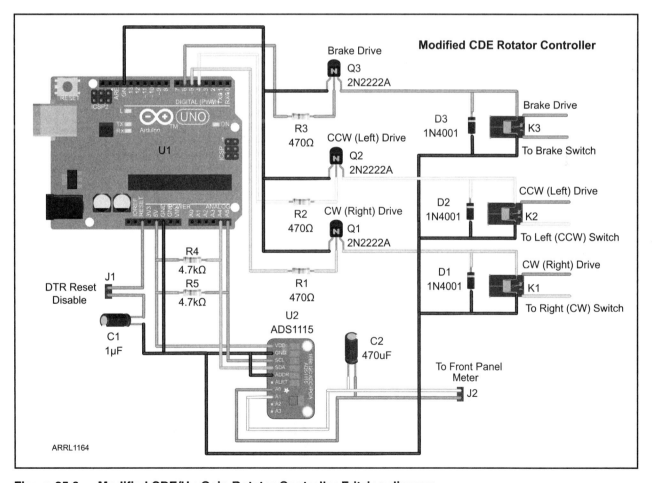

Figure 25.6 — Modified CDE/Hy-Gain Rotator Controller Fritzing diagram.

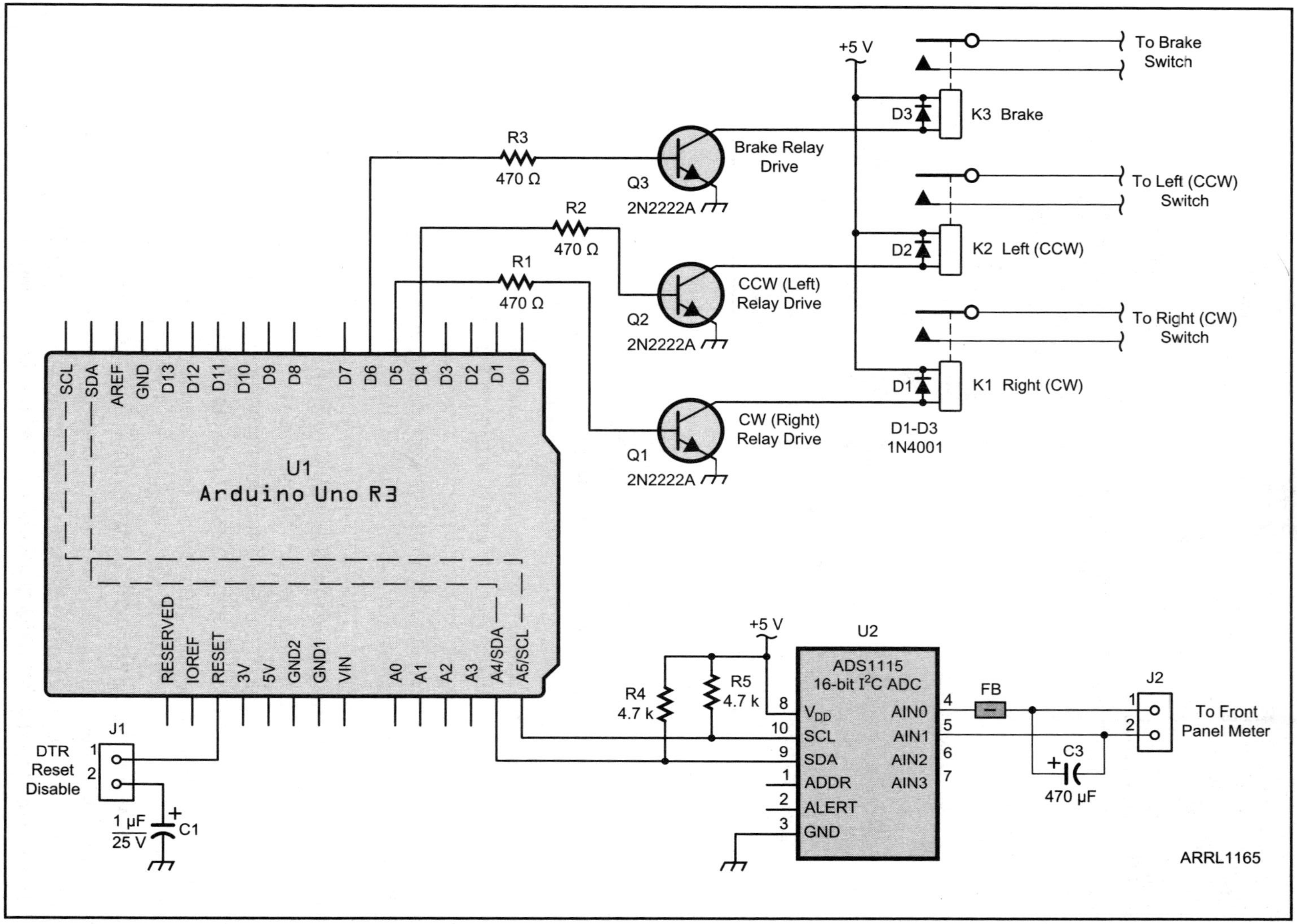

Figure 25.7 — Modified CDE/Hy-Gain Rotator Controller schematic diagram.

C1 — 1 µF, 16 V capacitor.
C2 — 470 µF, 16 V capacitor.
D1-D3 — 1N4001 diode.
FB — Ferrite bead choke.
J1 — 2 pin header jumper.
J2 — 2 pin header.
K1-K3 — 5 V SPST relay.
Q1-Q3 — 2N2222A transistor.
R1-R3 — 470 Ω resistor.
R4, R5 — 4.7 kΩ resistor.
U1 — Arduino Uno.
U2 — ADS1115 16-bit A/D converter module.

connectors to connect up the relays and rotator position sensing. **Figure 25.7** shows the schematic diagram for this project.

Finding a way to read the rotator position accurately with the A/D converter proved to be difficult. Reading the position sensor directly through a voltage divider to reduce the 12 V position sense voltage to 5 V brought the same noise issues that we had with the previous CDE/Hy-Gain rotator controller project. Searching through the rotator control box with a voltmeter and oscilloscope showed that the voltage across the front panel meter did not have as much ac noise, and we could have the A/D read the voltage across the analog meter differentially. By adding a 470 µF filtering capacitor across the meter, the ac hum was all but filtered out, and we had a usable position sense voltage. Since we're reading across the meter, the position sense voltage is very low (0 to 55 mV), so we'll need to set the ADS1115 gain to its most sensitive setting of 256 mV for a full scale reading. When you attach the wire coming from the analog meter, the meter lug closest to the center of the chassis is the higher potential (positive) side, which should be connected to A/D channel 0. The lower potential side (negative) is connected to A/D channel 1. This will allow the ADS1115 to read channel 0 and 1 differentially. Since we're reading differentially, the ac hum is no longer such a factor, and the input capacitor filters out whatever ac hum is remaining. A ferrite bead choke was also added to the position sense input to the A/D for added noise protection.

The relays are wired across the contacts on the motor and brake switches on the front panel. Be careful with the brake switch wiring. The brake switch actually controls the 120 V ac power to the 26 V ac power transformer, and the brake is released by applying power to the motor drive circuit. **Figure 25.8** shows the completed modifications to the CDE/Hy-Gain rotator control box.

Figure 25.8 — The finished controller modifications with all the wiring complete and the protoshield mounted on top of the Arduino Uno.

The Sketch

Now that the modifications to the CDE/Hy-Gain rotator control box are complete, we can start creating our sketch. You will see that the flowchart (**Figure 25.9**) for this project is very similar to the flowchart we used for the Yaesu Azimuth/Elevation Rotator Controller project. We'll use the EEPROM on the Arduino to save our azimuth calibration data. To allow interfacing with *Ham Radio Deluxe* for rotator control, we'll have the Arduino emulate the Yaesu

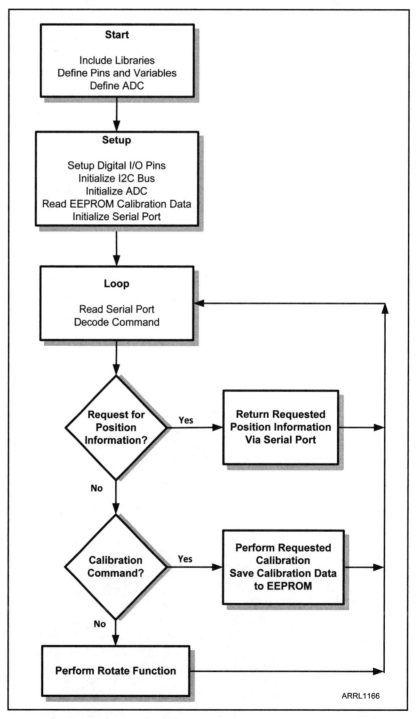

Figure 25.9 — Modified CDE/Hy-Gain Rotator Controller flowchart.

GS-232A in azimuth-only mode. This will allow us to use most of the same sketch we used for the Yaesu Azimuth/Elevation Rotator Controller project. All we have to do is make a few minor changes and remove the portions of the sketch that deal with elevation control. The Yaesu GS-232A command codes needed to communicate with *Ham Radio Deluxe* are listed in **Table 25.1**. Since the Yaesu GS-232A command set includes the commands needed to calibrate the zero and maximum azimuth values, we won't need the calibration switches we used in our previous CDE/Hy-Gain rotator controller project.

The complete sketch for the Modified CDE/Hy-Gain Rotator Controller can be found in **Appendix A** and online at **www.w5obm.us/Arduino**. Starting out with the sketch, we'll include the libraries we need, enable the debug mode for troubleshooting, and initialize the ADS1115 A/D converter object. The `ADS1115.h` and `I2Cdev.h` libraries have been customized to correct an error when handling 16 bit data. Be sure to use the modified libraries in your sketch. When your sketch is tested and debugged, don't forget to add the Reset Disable jumper on the protoshield and turn debug mode off, otherwise the *Ham Radio Deluxe* rotator controller software will not connect to the Arduino.

```
// Debug Mode must be off to communicate with Ham Radio Deluxe
#define debug_mode 1  // Set to 1 for debug data on Serial Port

#include <Wire.h>  // Include the I2C Communication Library
#include <EEPROM.h>  // Include the EEPROM Library
// Include the ADS1115.h Library (Library Updated to fix errors)
#include "ADS1115.h"
// Include I2CDev.h Library (Library Updated to fix errors)
#include "I2Cdev.h"

ADS1115 adc; // Define the ADS1115 as adc
```

The next thing we need to do is define the relay pins, the default positioning calibration data, and the speed on the serial port. You need to be sure whatever speed you choose to define matches the settings you plan to use in *Ham Radio Deluxe*. We'll also define the format for the calibration data that will be kept in the Arduino's onboard EEPROM.

Table 25.1
CDE/Hy-Gain Rotator Controller Commands
(Subset of Yaesu GS-232A Commands)

A	Stop Azimuth Rotation
C	Return Current Azimuth Value in Degrees (format +0aaa)
F	Set Azimuth Full Scale Calibration
L	Rotate Azimuth Counter-Clockwise
Maaa	Rotate Azimuth to aaa degrees
O	Set Azimuth Zero Calibration
R	Rotate Azimuth Clockwise
S	Stop All Rotation

```
#define rotate_left 4   // Define Rotate Left as Pin 4
#define rotate_right 5  // Define Rotate Right as Pin 5
#define brake 6   // Define the Brake pin

// Preset the Azimuth Zero Calibration Point to 0 if no EEPROM data
#define AZ_CAL_0_DEFAULT 0

// Preset the Azimuth Max Calibration Point to 7000 if no EEPROM data
#define AZ_CAL_MAX_DEFAULT 7000

#define AZ_Tolerance 2   // Set the Azimuth Accuracy Tolerance in degrees

#define BAUD_RATE 9600   // Set the Serial Port Baud rate to 9600

#define EEPROM_ID_BYTE 1   // EEPROM ID to validate EEPROM data location
#define EEPROM_ID  55   // EEPROM ID Value
#define EEPROM_AZ_CAL_0 2     // Azimuth Zero Calibration EEPROM location
#define EEPROM_AZ_CAL_MAX 4   // Azimuth Max Calibration Data EEPROM location
```

As the last step in the initialization portion of our sketch, we'll define all of the variables we'll be using in this sketch:

```
byte inByte = 0;   // incoming serial byte
byte serial_buffer[50];   // incoming serial byte buffer

// The index pointer variable for the Serial buffer
int serial_buffer_index = 0;

int set_AZ;   // Azimuth set value
int current_AZ;   // Current Azimuth raw value

String Serial_Send_Data; // Data to send to Serial Port
int AZ_0;   // Azimuth Zero Value from EEPROM
int AZ_MAX; // Azimuth Max Value from EEPROM

int AZ_Degrees; // mapped AZ ADC value to Degrees
String Requested_AZ; // RS232 Requested Azimuth - M command
int AZ_To; // Requested AZ Move
int AZ_Distance; // Distance to move AZ
```

In the `setup()` loop, we'll start out by setting the digital I/O pin modes for the brake solenoid and motor drive relays. Then we'll turn off the relays to make sure everything is turned off before we begin.

```
// Define the Control Pins as Outputs
pinMode(rotate_left, OUTPUT); // define the rotate left relay pin
pinMode(rotate_right, OUTPUT);  // define the rotate right relay pin
pinMode(brake, OUTPUT);   // define the brake solenoid relay pin
```

```
// Turn off all the relays just to be sure
digitalWrite(rotate_left, LOW);  // Turn off the rotate left relay
digitalWrite(rotate_right, LOW);  // Turn off the rotate right relay
digitalWrite(brake, LOW);  // Turn off the brake solenoid relay
```

Next, we'll start the serial port, the I²C bus, and the ADS1115. We'll then configure the A/D to run continuously at 32 samples per second. We'll also set the A/D gain to 256 mV and have the A/D read differentially between input channels 0 and 1.

```
Serial.begin(BAUD_RATE); // initialize serial communication

Wire.begin();  // join I2C bus

adc.initialize(); // initialize ADS1115 16 bit A/D chip

Wire.beginTransmission(0x48); // Begin direct ADC communication
//  Connect to adc and send two bytes - Set Config Reg to all Ones
Wire.write(0x1);
Wire.write(0x7F);   //  MSB
Wire.write(0xFF);    //  LSB
Wire.endTransmission();   // End the direct ADC Communication

// Set the ADC to free running conversion mode
adc.setMode(ADS1115_MODE_CONTINUOUS);

// set the ADC gain to 0.256 Volt range, 0.007813 Volts/step
adc.setGain(ADS1115_PGA_0P256);

// set ADC sample rate to 32 samples per second
adc.setRate(ADS1115_RATE_32);

// Set the ADC to AN0+ AN1 Differential Mode
adc.setMultiplexer(ADS1115_MUX_P0_N1);
```

In the last portion of the `setup()` loop, we'll turn off the flag we use to indicate rotator movement and read the rotator position calibration data from the Arduino's EEPROM. If there are no calibration values saved, the sketch will use the defined default calibration values.

```
set_AZ = -1;  // Preset the Azimuth Move Variables

// Read the Azimuth Calibration Values from EEPROM
read_eeprom_cal_data();
```

As with the Yaesu Azimuth/Elevation Rotator Controller project, there are only two statements in the main `loop()`. The major portion of the work is done by function calls. The `check_serial()` function checks the serial port for incoming commands from *Ham Radio Deluxe* and executes any commands it receives. The `check_move()` function takes care of the actual rotator movement.

```
check_serial();  // Check the Serial Port for Data
check_move();    // Check to see if executing move command
```

The functions in this sketch handle most of the work. The first function, `read_eeprom_cal_data()`, reads the calibration data stored in the Arduino's onboard EEPROM and places it into the variables we use to map the raw rotator position data into the actual azimuth in degrees. If there is no valid calibration data, the default values are used and also saved in the Arduino's EEPROM.

```
void read_eeprom_cal_data()  // Function to Read the Azimuth Calibration Data
{
// Verify the EEPROM has valid data
if (EEPROM.read(EEPROM_ID_BYTE) == EEPROM_ID)
  {
    if (debug_mode) // If in Debug Mode Print the Calibration Values
    {
      Serial.println("Read EEPROM Calibration Data Valid ID");
      Serial.println((EEPROM.read(EEPROM_AZ_CAL_0) * 256) + EEPROM.read(EEPROM_AZ_CAL_0 + 1),DEC);
      Serial.println((EEPROM.read(EEPROM_AZ_CAL_MAX) * 256) + EEPROM.read(EEPROM_AZ_CAL_MAX + 1),DEC);
    }

    // Read the Azimuth Zero Calibration Value from EEPROM
    AZ_0 = (EEPROM.read(EEPROM_AZ_CAL_0)*256) + EEPROM.read(EEPROM_AZ_CAL_0 + 1);
    // Read the Azimuth Maximum Calibration Value from EEPROM
    AZ_MAX = (EEPROM.read(EEPROM_AZ_CAL_MAX)*256) + EEPROM.read(EEPROM_AZ_CAL_MAX + 1);

  } else {  // initialize eeprom to default values
  if (debug_mode)
  {
   Serial.println("Read EEPROM Calibration Data Invalid ID - setting to defaults");
  }
  AZ_0 = AZ_CAL_0_DEFAULT;   // Set the Calibration To Default Values
  AZ_MAX = AZ_CAL_MAX_DEFAULT;
  write_eeprom_cal_data();   // Write the Default Values to EEPROM
  }
}
```

The `write_eeprom_cal_data` function is used to write the calibration data to the Arduino's onboard EEPROM:

```
// Function to Write the Calibration Values to EEPROM
void write_eeprom_cal_data()
{
  Serial.println("Writing EEPROM Calibration Data");

  EEPROM.write(EEPROM_ID_BYTE,EEPROM_ID);  //   Write the EEPROM ID
  // Write the Azimuth Zero Calibration High Order Byte
  EEPROM.write(EEPROM_AZ_CAL_0,highByte(AZ_0));
  // Write the Azimuth Zero Calibration Low Order Byte
  EEPROM.write(EEPROM_AZ_CAL_0 + 1,lowByte(AZ_0));
  // Write the Azimuth Max Calibration High Order Byte
  EEPROM.write(EEPROM_AZ_CAL_MAX,highByte(AZ_MAX));
  // Write the Azimuth Max Calibration Low Order Byte
  EEPROM.write(EEPROM_AZ_CAL_MAX + 1,lowByte(AZ_MAX));
}
```

The `check_serial()` function will read the incoming characters from the PC, then decode and execute the Yaesu GS-232A command. If you are not using *Ham Radio Deluxe* to control your rotator, you may need to uncomment the statement that echoes the received data back to the PC. *Ham Radio Deluxe* will not connect to the rotator controller if debug mode is on or if the received characters are echoed back to the PC.

```
void check_serial() // Function to check for data on the Serial port
{
  if (Serial.available() > 0) // Get the Serial Data if available
  {
    inByte = Serial.read();  // Get the Serial Data

  // You may need to uncomment the following line if your PC software
  // will not communicate properly with the controller
  //   Serial.print(char(inByte));  // Echo back to the PC

    if (inByte == 10)  // ignore Line Feeds
    {
    return;
    }
    if (inByte !=13)  // Add to buffer if not CR
    {
      serial_buffer[serial_buffer_index] = inByte;

      if (debug_mode) // Print the Character received if in Debug mode
      {
        Serial.print("Received = ");
        Serial.println(serial_buffer[serial_buffer_index]);
      }
```

```
      serial_buffer_index++;  // Increment the Serial Buffer pointer

} else {  // It's a Carriage Return, execute command

//If first character of command is lowercase, convert to uppercase
if ((serial_buffer[0] > 96) && (serial_buffer[0] < 123))
{
   serial_buffer[0] = serial_buffer[0] - 32;
}
```

Once the command has been received, we'll use a `switch...case()` statement to decode and execute the command:

```
switch (serial_buffer[0]) {  // Decode first character of command

    case 65:   // A Command - Stop the Azimuth Rotation

      if (debug_mode) {Serial.println("A Command Received");}
      az_rotate_stop(); // call the rotate stop function
      break;

    case 67:       // C - return current azimuth

      if (debug_mode)    // Return the Buffer Index Pointer in Debug Mode
      {
        Serial.println("C Command Received");
        Serial.println(serial_buffer_index);
      }
      send_current_az();  // Return Azimuth in degrees
      break;

    case 70:   // F - Set the Max Calibration

      if (debug_mode)
      {
        Serial.println("F Command Received");
        Serial.println(serial_buffer_index);
      }
      set_max_az_cal();  // F - Set the Max Azimuth Calibration
      break;

    case 76:   // L - Rotate Azimuth CCW

      if (debug_mode)
      {
        Serial.println("L Command Received");
      }
      rotate_az_ccw();   // Call the Rotate Azimuth Left (CCW) Function
      break;
```

```
    case 77:  // M - Rotate to Set Point

      if (debug_mode)
      {
        Serial.println("M Command Received");
      }
      rotate_to();  // Call the Rotate to Set Point Command
      break;

    case 79:  // O - Set Zero Calibration

      if (debug_mode)
      {
        Serial.println("O Command Received");
        Serial.println(serial_buffer_index);
      }
      set_0_az_cal();  // O - Set the Azimuth Zero Calibration
      break;

    case 82:  // R - Rotate Azimuth CW

      if (debug_mode)
      {
        Serial.println("R Command Received");
      }
      rotate_az_cw();  // Call the Rotate Azimuth Right (CW) Function
      break;

    case 83:  // S - Stop All Rotation

      if (debug_mode)
      {
        Serial.println("S Command Received");
      }
      az_rotate_stop();  // Call the Stop Azimith Rotation Function
      break;

  }
```

After the decoded command has been executed, the serial data buffer is cleared and the sketch is ready to receive the next command:

```
// Clear the Serial Buffer and Reset the Buffer Index Pointer
serial_buffer_index = 0;
serial_buffer[0] = 0;
```

The `send_current_az()` function will read the A/D converter and convert the raw A/D value to a calibrated azimuth value. The azimuth data is adjusted to allow for the 0°/360° position being at midscale instead of at the minimum and maximum values. The azimuth value in degrees is then sent to the PC via the USB port.

```
void send_current_az() // Send the Current Azimuth Function
{
  read_adc();   // Read the ADC

  // Map Azimuth to degrees
  if (debug_mode)
  {
    Serial.println(current_AZ);
  }
  // Map the Current Azimuth to Degrees
  AZ_Degrees = map(current_AZ, AZ_0, AZ_MAX, 0, 360);

  // Correction Since Azimuth Reading starts at Meter Center Point
  if (AZ_Degrees > 180)
  {
    AZ_Degrees = AZ_Degrees - 180;
  } else {
    AZ_Degrees = AZ_Degrees + 180;
  }
  if (debug_mode)
  {
    Serial.println(AZ_Degrees);
  }
  // Send it back via serial
  Serial_Send_Data = "";
  if (AZ_Degrees < 100)  // pad with 0's if needed
  {
    Serial_Send_Data = "0";
  }
  if (AZ_Degrees < 10)
  {
    Serial_Send_Data = "00";
  }
  // Send the Azimuth in Degrees
  Serial_Send_Data = "+0" + Serial_Send_Data + String(AZ_Degrees);
  Serial.println(Serial_Send_Data);   // Return value via USB port
}
```

The set_max_az_cal() function is used to set the calibration value for the maximum azimuth rotation. This value is then saved to the AZ_MAX calibration variable, and to the Arduino's onboard EEPROM.

```
void set_max_az_cal() // Set the Max Azimuth Calibration Function
{
  Serial.println("Calibrate Max AZ Function");

  read_adc();   // Read the ADC

  // save current az and el values to EEPROM - Zero Calibration
  Serial.println(current_AZ);

  // Set the Azimuth Maximum Calibration to Current Azimuth Reading
  AZ_MAX = current_AZ;

  write_eeprom_cal_data();  // Write the Calibration Data to EEPROM
  Serial.println("Max Azimuth Calibration Complete");
}
```

The `rotate_az_cw()` and `rotate_az_ccw()` functions are used to control the brake solenoid and the rotator motor drive relays:

```
void rotate_az_ccw() // Function to Rotate Azimuth Left (CCW)
{
  digitalWrite(brake, HIGH);  // Release the brake
  digitalWrite(rotate_left, HIGH);  // Set the Rotate Left Pin High
  digitalWrite(rotate_right, LOW);  // Make sure the Rotate Right Pin is Low
}

void rotate_az_cw() // Function to Rotate Azimuth Right (CW)
{
  digitalWrite(brake, HIGH);  // Release the brake
  digitalWrite(rotate_right, HIGH);   // Set the Rotate Right Pin High
  digitalWrite(rotate_left, LOW);    // Make sure the Rotate Left Pin Low
}
```

The `az_rotate_stop()` function is used to stop all rotation. The motor drive relays are de-energized, and a one second delay is added to allow the rotator and antenna to coast to a stop. After the one second delay expires, the brake solenoid is de-energized and the brake is engaged.

```
void az_rotate_stop() // Function to Stop Azimuth Rotation
{
  digitalWrite(rotate_right, LOW);  // Turn off the Rotate Right Pin
  digitalWrite(rotate_left, LOW);   // Turn off the Rotate Left Pin
  set_AZ = -1;
  delay(1000);
  digitalWrite(brake, LOW);  // Engage the brake
}
```

The `rotate_to()` function is used to move the rotator to the commanded position. The rotator will turn automatically until it reaches the desired position or is given a Stop command.

```
void rotate_to() // Function to Rotate to Set Point
{
  if (debug_mode)
  {
    Serial.println("M Command -  Rotate Azimuth To Function");
  }
  // Decode Command - Format Mxxx where xxx = Degrees to Move to
  if (debug_mode)
  {
    Serial.println(serial_buffer_index);
  }
  if (serial_buffer_index == 4)  // Verify the Command is the proper length
  {
    if (debug_mode)
    {
      Serial.println("Value in [1] to [3]?");
    }
    // Decode the Azimuth Value
    Requested_AZ = (String(char(serial_buffer[1])) + String(char(serial_buffer[2])) + String(char(serial_buffer[3]))) ;

    AZ_To = (Requested_AZ.toInt()); // AZ Degrees to Move to as an integer
    if (AZ_To <0) // Make sure we don't go below 0 degrees
    {
      AZ_To = 0;
    }
    if (AZ_To >360) // Make sure we don't go over 360 degrees
    {
      AZ_To = 360;
    }
    if (AZ_To > 180) // Adjust for Meter starting at midscale
    {
      AZ_To = AZ_To - 180;
    } else {
      AZ_To = AZ_To + 180;
    }
    if (debug_mode)
    {
      Serial.println(Requested_AZ);
      Serial.println(AZ_To);
    }

    // set the move flag and start
    read_adc();  // Read the ADC

    // Map it to degrees
    if (debug_mode)
    {
```

```
      Serial.println(current_AZ);
    }
    // Map the Azimuth Value to Degrees
    AZ_Degrees = map(current_AZ, AZ_0, AZ_MAX, 0, 360);
    if (debug_mode)
    {
      Serial.println(AZ_Degrees);
    }
    AZ_Distance = AZ_To - AZ_Degrees;  // Figure out far we have to move
    set_AZ = AZ_To;
    // No move needed if we're within the defined tolerance range
    if (abs(AZ_Distance) <= AZ_Tolerance)
    {
      az_rotate_stop();  // Stop the Azimuth Rotation
      set_AZ = -1;  // Turn off the Move Command
    } else {  // Move Azimuth - figure out which way
      if (AZ_Distance > 0)   //We need to move right (CW)
      {
        rotate_az_cw();  // If the distance is positive, move right (CW)
      } else {
        rotate_az_ccw();  // Otherwise, move left (CCW)
      }
    }
  }
}
```

The `set_0_az_cal()` function is used to set the azimuth zero calibration value. This value is then stored to the `AZ_0` calibration variable and saved to the Arduino's onboard EEPROM.

```
void set_0_az_cal() // Set Azimuth Zero Calibration
{
  Serial.println("Calibrate Zero Function");

  read_adc();  // Read the ADC

  // save current Azimuth value to EEPROM - Zero Calibration
  Serial.println(current_AZ);

  AZ_0 = current_AZ;  // Set the Azimuth Zero Calibration to current position
  write_eeprom_cal_data();  // Write the Calibration Data to EEPROM
  Serial.println("Zero Azimuth Calibration Complete");
}
```

The `read_adc()` function is used to read the ADS1115 module and returns the raw uncalibrated rotator position digital value:

```
void read_adc() // Function to read the ADC
{
 if (debug_mode)
 {
   Serial.println("Read ADC Function  ");
 }

  int RotorValue;  // Variable to store the rotor value
  adc.setRate(ADS1115_RATE_32); // Set the ADC rate to 32 samples/sec
  adc.setGain(ADS1115_PGA_0P256); // Set the ADC gain to 0.007813 Volts

// Set the ADC to Channel 0 AN0+ AN1 Differential Mode
  adc.setMultiplexer(ADS1115_MUX_P0_N1);
  delay(100); // adc settling delay

  // Read ADC Channel 0 and 1 Differentially
  current_AZ = adc.getDifferential();
}
```

Finally, the `check_move()` function handles the actual rotation. If the sketch has received a move command, the `check_move()` function will manage the rotation until the commanded rotator position has been reached or a Stop command is received.

```
void check_move() // Check to see if we've been commanded to move
{
  if (set_AZ != -1) {   // We're moving - check and stop as needed
    read_adc();  // Read the ADC
    // Map AZ to degrees
    if (debug_mode)
    {
      Serial.println(current_AZ);
    }
    // Map the Current Azimuth reading to Degrees
    AZ_Degrees = map(current_AZ, AZ_0, AZ_MAX, 0, 360);

    if (debug_mode)
    {
      Serial.println(AZ_Degrees);
    }

    if (set_AZ != -1) // If Azimuth is moving
    {
      AZ_Distance = set_AZ - AZ_Degrees;  // Check how far we have to move
      // No move needed if we're within the tolerance range
```

```
      if (abs(AZ_Distance) <= AZ_Tolerance)
      {
        az_rotate_stop();  // Stop the Azimuth Rotation
        set_AZ = -1;  // Turn off the Azimuth Move Command
      } else {  // Move Azimuth - figure out which way
        if (AZ_Distance > 0)   //We need to move right (CW)
        {
          rotate_az_cw();  // Rotate right (CW) if positive
        } else {
          rotate_az_ccw();  // Rotate left (CCW) if negative
        }
      }
    }
  }
}
```

You can see how much of the Yaesu Azimuth/Elevation Controller sketch that we were able to re-use. This is one reason why working with the Arduino is so much fun. Once you have built your own Arduino sketch and library collection, you can re-use them endlessly in your future projects, greatly reducing your sketch development time. Being able to re-use huge blocks of code made this a very easy sketch to create, since all that we had to do was a little modification here and there. Once you have the sketch ready for testing and debugging, you can use the Serial Monitor to send the Yaesu GS-232A commands to the rotator controller.

Once you have everything debugged and working, don't forget to upload the sketch with the debug mode turned off, and the Reset Disable jumper installed on the Arduino protoshield, before trying to control the rotator with *Ham Radio Deluxe*.

Now we're ready to have some real fun. With the sketch complete, we're ready to test the interface and set the azimuth calibration values. For this test, you can use the Arduino IDE's Serial Monitor to send the Yaesu GS-232A commands directly to the Arduino. Manually rotate the azimuth to zero (the front panel analog meter at minimum deflection) and use the "O" command to set the zero calibration point. Then, manually rotate the azimuth to full-scale on the front panel analog meter and use the "F" command to set the azimuth full-scale value. Once you have done that, your controller is calibrated, and the calibration values are stored in the Arduino's onboard EEPROM. You can recalibrate your controller at any time by repeating the calibration process. You can now send the various Yaesu GS-232A controller commands listed back in Table 25.1 to test all the functions on your controller.

Now that your controller is calibrated, you're ready to use the controller with the *Ham Radio Deluxe* rotator controller software on your PC. Don't forget to install the Reset Disable jumper and have the sketch loaded with the debug mode turned off. Configure the *Ham Radio Deluxe* rotator controller to use the Arduino's COM port, set the baud rate to match the serial port speed setting in the sketch, and select the Yaesu GS-232A/Az rotator controller in *Ham Radio Deluxe*. Now your rotator can be controlled by *Ham Radio Deluxe*, and the

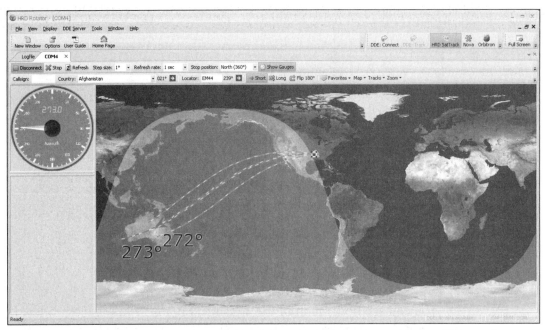

Figure 25.10 — Screen shot of *Ham Radio Deluxe* controlling the modified CDE/Hy-Gain rotator controller.

current rotator position will be displayed on the *HRD* rotator controller screen as shown in **Figure 25.10**. If the rotator turns correctly, but *Ham Radio Deluxe* doesn't show anything on the rotator dial, you need to verify that the Arduino does not have the command echo statement uncommented in the `check_serial()` function. *Ham Radio Deluxe* doesn't like it when you echo the command back to it. If everything went well, you now have a fully operational CDE/Hy-Gain rotator controller that can be controlled by *Ham Radio Deluxe* or other software on your PC.

Enhancement Ideas

Because this needed to be a fully functional project, there's not a whole lot left for you to enhance. You could use an Arduino Nano and shrink everything to a perfboard mounted inside the CDE/Hy-Gain rotator control box, or you could even put the Arduino in a small external enclosure and run the rotator sense and relay control wires through a small hole drilled in the rotator controller chassis to save you the fun of having to cut out a square hole in the rotator controller chassis for the Arduino's USB connector. You might also want to add an RGB LED to the front of your CDE/Hy-Gain control box so you can tell when the PC is controlling the rotator. One final enhancement would be to add an infrared LED detector and use an infrared controller to control your rotator wirelessly. This would allow you to move your rotator control box off your desk, and clear up some space for more stuff.

References

Ham Radio Deluxe — **www.ham-radio-deluxe.com**
Texas Instruments — **www.ti.com**
Yaesu — **www.yaesu.com**

CHAPTER 26

In Conclusion

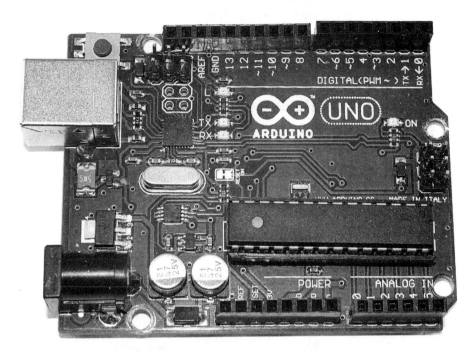

The previous chapters in this book have shown many ways to use the Arduino Uno to enhance your ham radio capabilities. Look at these projects as the beginning of your Arduino experience.

In this book, we've only been able to scratch the surface of what the Arduino can do. There are so many more shields and modules that can be used in ham radio projects that we haven't gotten to play with yet, but that too is part of the fun of the Arduino. Like the Erector Set from my childhood, there is no end to the things that can be built, especially with new add-ons coming out all the time. In this book, I have tried to cover the broad spectrum of ham radio, creating projects that briefly touched as many aspects as possible. You know as well as I do that can't be done entirely in a single book. This is where you come in. Hopefully, this book has given you the spark of inspiration and knowledge you need to go out and create your own projects.

If you need to learn more about the Arduino and what it can do, there are several websites that have excellent tutorials and projects. The Arduino Playground, Instructables.com, SparkFun.com and Adafruit.com are just several that come to mind, but there are many others to choose from.

So where do you start? Well, how about starting with the things that are not in this book. For example, the Arduino can be web-enabled with the Ethernet shield. With the Ethernet shield, you can create Arduino projects that can be accessed via the Internet, and control and monitor your shack remotely.

Another area that could prove quite interesting is the linking of an Android phone to the Arduino. There are numerous Android apps available that allow your Arduino to communicate with an Android device via Bluetooth. Now you can create telemetry and control applications for the Arduino. To help you along this path, take a look at Android apps such as *ArduDroid*, *ArduinoCommander*, *Arduino Uno Communicator*, and *Arduino Total Control*, just to name a few. For the PC side, there are programs such as *Processing* and *Ardulink* that will allow you to create applications on your PC that can communicate with the Arduino.

Voice Recognition

While I was wrapping up this book, I bought an EasyVR voice recognition shield shown in **Figure 26.1**. This shield can respond to 28 speaker-independent, and 32 speaker-specific user-defined commands. It comes complete with a software library for the Arduino, so how hard can it be to get this new toy working? I doubt it will be very hard at all. We've had voice-operated (VOX) transmitters for years, but now, we can take that to a whole new level and make our entire shack voice-controlled. I am sitting here thinking my usual phrase, "Wouldn't it be cool if you could just sit back, give your antenna rotator a voice command, and have it turn your antennas for

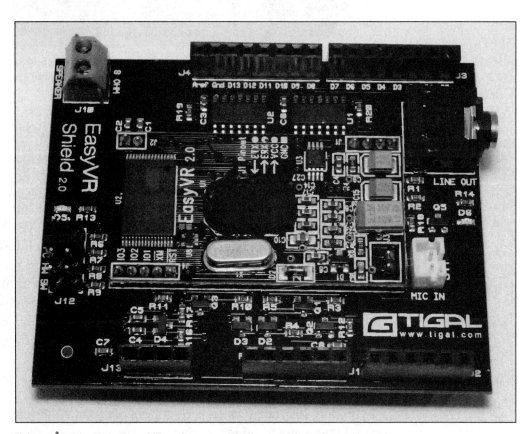

Figure 26.1 — The EasyVR voice recognition shield.

you automatically?" Why stop there? You can connect an Arduino to your transceiver's computer interface and do everything by voice commands. For me, the possibilities are endless with this new shield.

With motion and infrared sensors, you could use your Arduino to turn on everything in your shack when you walk in, and turn everything off when you leave. Sure you can use a plain old power switch, but that's just not that same as creating your own Arduino project to do it for you.

And we've only scratched the surface of what can be done with the direct digital frequency synthesis (DDS) modules. If you link a DDS module with an SWR sensing unit, you can build your own antenna analyzer. If you link that to your PC or Android phone, or use one of the color TFT displays on the Arduino itself, you can graph the entire SWR curve for your antennas. Going further, you can use a DDS module and create your own Arduino-based transceiver.

TEN-TEC Rebel Open Source Transceiver

If you're not into designing your own transceiver, but still want to experiment with one, there are Open Source products such as the TEN-TEC Rebel Model 506 QRP CW transceiver shown in **Figures 26.2** and **26.3**. The Rebel is controlled by a Digilent chipKIT Uno32, which is a more powerful software and hardware-compatible variant of the Arduino Uno. Because the Rebel is Open Source, both hardware and software — everything you need to roll your own extra features and enhancements — is provided to you under the Open Source umbrella. The TEN-TEC Rebel Yahoo User's group has already created a number of enhancements to the Rebel, including various displays, CW keyers, and more, all of which are shared under Open Source licensing for all to use and enjoy. I feel that the Rebel is just the start of an Open Source revolution in ham radio, and it's being fueled by the Arduino and its cousins.

The world of Open Source is truly a wide-ranging and wonderful world. It's like having thousands of mentors and fellow Arduino developers just a mouse click away, freely sharing their knowledge, creations, and questions with you. This book could never have happened had others not shared their knowledge and creations for me to learn from and to build upon.

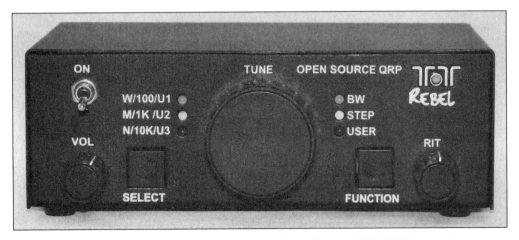

Figure 26.2 — The TEN-TEC Rebel Model 506 Open Source QRP CW transceiver.

Figure 26.3 — Inside view of the TEN-TEC Rebel showing an Arduino protoshield mounted on the chipKit Uno 32 shield expansion pins.

So, as you go out and start creating your own magic with the Arduino, please remember to share back to the Open Source community so that others can follow in your footsteps. And by all means, please feel free to share with me what you have done, both with the projects in this book and the projects you create on your own. Who knows, you may end up creating that one enhancement that I would love to have and didn't even think about, or provided the spark for my next Arduino adventure.

73, Glen Popiel, KW5GP

References

Adafruit Industries — **www.adafruit.com**
Android Apps — **play.google.com/store/apps**
ArduinoCommander — **arduinocommander.blogspot.com**
Arduino-Communicator — **github.com/jeppsson/Arduino-Communicator**
Arduino Playground — **playground.arduino.cc**
Ardulink — **www.ardulink.org**
Instructables — **www.instructables.com**
Processing — **www.processing.org**
SparkFun Electronics — **www.sparkfun.com**
TEN-TEC — **www.tentec.com**
TEN-TEC Rebel User's Group — **groups.yahoo.com/neo/groups/ TenTec506Rebel**
VeeaR EasyVR — **www.veear.eu**

APPENDIX A

Sketches and Libraries

This is a list of all of the sketches and libraries used to create the projects in this book. You can download a PDF file with the complete sketches and libraries from the ARRL website at **www.arrl.org/arduino**. You can also download the sketch and library files themselves from **www.w5obm.us/Arduino**. Links to the original libraries are also provided to allow you to download the current version of the libraries if desired.

Chapter 7 — Random Code Practice Generator

Libraries Required:
  LiquidCrystal_I2C
  Morse (customized)
Sketch Required:
  Random_Code_Oscillator.ino

Chapter 8 — CW Beacon and Foxhunt Keyer

Libraries Required:
  Morse (customized)
Sketch Required:
  CW_Beacon.ino

Chapter 9 — Fan Speed Controller

Libraries Required:
  OneWire
Sketch Required:
  Fan_Speed_Controller.ino

Chapter 10 — Digital Compass

Libraries Required:
  LiquidCrystal_I2C
  HMC5883L
Sketch Required:
  Digital_Compass.ino

Chapter 11 — Weather Station

Libraries Required:
  dht
  LCD5110_Basic
Sketch Required:
  Weather_Station.ino

Chapter 12 — RF Probe with LED Bar Graph

Libraries Required:
  None
Sketch Required:
  RF_Probe.ino

Chapter 13 — Solar Battery Charging Monitor

Libraries Required:
  None
Sketch Required:
  Solar_Charging_Monitor.ino

Chapter 14 — On Air Indicator

Libraries Required:
  None
Sketch Required:
  On_Air_Indicator.ino

Chapter 15 — Talking SWR Meter

Libraries Required:
  LCD5110_Basic
Sketch Required:
  Talking_SWR_Meter.ino

Chapter 16 — Talking GPS/UTC Time/Grid Square Indicator

Libraries Required:
  LCD5110_Basic
  TinyGPS
Sketch Required:
  Grid_Square_Display.ino

Chapter 17 — Iambic Keyer

Libraries Required:
  LCD5110_Basic
Sketch Required:
  Iambic_Keyer.ino

Chapter 18 — Waveform Generator

Libraries Required:
   LiquidCrystal_I2C
Sketch Required:
   Waveform_Generator.ino

Chapter 19 — PS/2 CW Keyboard

Libraries Required:
   I2C
   LiquidCrystal_I2C
   Morse (customized)
   PS2Keyboard (customized)
Sketch Required:
   CW_Memory_Keyer.ino

Chapter 20 — Field Day Satellite Tracker

Libraries Required:
   None
Sketch Required:
   Satellite_Tracker.ino

Chapter 21 — Azimuth/Elevation Rotator Controller

Libraries Required:
   ADS1115 (customized)
   I2Cdev (customized)
Sketch Required:
   Yeasu_GS_232A_Rotor_Controller.ino

Chapter 22 — CW Decoder

Libraries Required:
   LiquidCrystal_I2C
   MorseEnDecoder
Sketch Required:
   CW_Decoder.ino

Chapter 23 — Lightning Detector

Libraries Required:
   AS3935
   I2C
   LCD5110_Basic
Sketch Required:
   Lightning_Detector.ino

Chapter 24 — CDE/Hy-Gain Rotator Controller

Libraries Required:
ADS1115 (customized)
I2Cdev (customized)
LiquidCrystal_I2C
Timer

Sketch Required:
Rotor_Controller.ino

Chapter 25 — Modified CDE/Hy-Gain Rotator Controller

Libraries Required:
ADS1115 (customized)
I2Cdev (customized)

Sketch Required:
Modified_CDE_Rotor_Controller.ino

Libraries

All libraries used for the projects in this book can be downloaded either from the link provided or from **www.w5obm.us/Arduino**. In the case of the `LCD5110_Basic` library used for the Nokia 5110 display, the library developer has requested that you download the library directly from their website to ensure the latest version.

Library Links

LiquidCrystal_I2C — **hmario.home.xs4all.nl/arduino/LiquidCrystal_I2C**
OneWire — **www.pjrc.com/teensy/td_libs_OneWire.html**
HMC5883L — **bildr.org/2012/02/hmc5883l_arduino**
dht — **arduino.cc/playground/Main/DHTLib**
LCD5110_Basic — **www.henningkarlsen.com/electronics/library.php?id=44**
TinyGPS — **www.arduiniana.org**
I2C — **github.com/rambo/I2C**
MorseEnDecoder — **code.google.com/p/morse-endecoder**
AS3935 — **www.github.com/SloMusti/AS3935-Arduino-Library**
Timer — **www.github.com/JChristensen/Timer**

All customized libraries available at **www.w5obm.us/Arduino**:
Morse
PS2Keyboard
ADS1115
I2Cdev

APPENDIX B

Design and Schematic Tools

There are two primary tools that I use to design and document my Arduino projects, *Fritzing* and CadSoft *EaglePCB*. Documenting your projects is very important. Sometimes you'll revisit a project that you built months ago (or longer), and need to be able to pick up where you left off. Without good documentation, you'll lose a lot of time trying to remember how and why you did something the way it was done.

Fritzing

Fritzing is a free Open Source design tool I use to produce drawings of how the circuit will look on a breadboard. Fritzing produces a realistic-looking layout of the breadboard design and all you have to do is match your actual breadboard wiring to the Fritzing drawing to construct your prototype.

Fritzing comes with a parts library that contains many of the components you will be using to design your projects. Fritzing will run on *Windows*, Mac *OS X*, and *Linux*. New parts definitions are constantly being added to the Fritzing distribution and their website for download. The parts in Fritzing are customizable, and you can modify existing parts templates, or create your own. The Fritzing website, **www.fritzing.org**, has a series of excellent tutorials to help you learn how to use Fritzing in your circuit designs. *Fritzing* can also be used to draw schematics and create the actual printed circuit board patterns for etching circuit boards of your finished projects.

Inkscape

Fritzing graphics for new parts are easily created using another free Open Source program called *Inkscape*. Inkscape is a free, Open Source graphics editor that can create the scalable vector graphics (SVG) files used by Fritzing. Inkscape will run on *Windows*, Mac *OS X*, and *Linux*. The Inkscape website, **www.inkscape.org**, has a number of excellent tutorials to help you along your way when you need to design your own parts for Fritzing.

EaglePCB

Cadsoft's *EaglePCB* is my tool of choice for creating schematic drawings. Eagle is used commercially by many companies to produce schematics and printed circuit boards. EaglePCB will run on *Windows*, Mac *OS X*, and *Linux*. As with Fritzing, EaglePCB comes with an extensive library of components. If the component you need is not in a library, you can easily create or download a new component, or modify an existing one to get what you need. EaglePCB will also produce the Gerber data files used to create etched circuit boards. While Fritzing does an excellent job creating the breadboard layouts, I prefer to use EaglePCB to create my finished schematic diagrams.

EaglePCB has several levels of licensing. The freeware Eagle Light Edition will do just about everything the average hobbyist needs. The Eagle Light Edition limits you to a circuit board size of 100 × 80 mm (4 × 3.2 inches), two signal layers and a single design sheet. The Light Edition also limits you to a single user and nonprofit applications. If you are planning to create circuit boards for your projects, the layout and autorouting features can be added to the Light Edition for $69.

The Eagle Hobbyist version allows you to create circuit boards up to 160 by 100 mm (6.3 × 3.9 inches), six signal layers and up to 99 design sheets. The Hobbyist version also includes the layout and autorouting features. A single-user license for the Eagle Hobbyist version currently costs $169 and you are restricted to noncommercial use.

If you plan to sell your finished products and designs, you can purchase either the Eagle Standard or Professional versions starting at $315 for a single-user license.

As with Fritzing and Inkscape, the CadSoft website (**www.cadsoftusa.com**) has an excellent series of tutorials and videos to help you learn how to use EaglePCB to create your schematic diagrams.

APPENDIX C

Vendor Links and References

The following companies offer parts and supplies of general interest to Arduino experimenters. Many of their products were used in projects described in this book.

4D Systems — www.4dsystems.com.au
Adafruit Industries — www.adafruit.com
Amazon — www.amazon.com
Austriamicrosystems AG — www.austriamicrosystems.com
Crisp Concept — www.crispconcept.com
DFRobot — www.dfrobot.com
Diligent — www.digilentinc.com
eBay — www.ebay.com
Embedded Adventures — www.embeddedadventures.com
MFJ — www.mfjenterprises.com
Midnight Design Solutions — www.midnightdesignsolutions.com
Pololu Robotics and Electronics — www.pololu.com
RadioShack — www.radioshack.com
Smarthome — www.smarthome.com
Solarbotics — www.solarbotics.com
SparkFun Electronics — www.sparkfun.com
TEN-TEC — www.tentec.com
Tindie — www.tindie.com
West Mountain Radio — www.westmountainradio.com
X10 — www.x10.com
Yaesu — www.yaesu.com
ZiGo — www.zigo.am

NOTES

NOTES

NOTES

NOTES

INDEX

Note: The letters "ff" after a page number indicate coverage of the indexed topic on succeeding pages.

A

AD9833 programmable waveform generator:....... 18-4
AD98xx series DDS module: 3-21
ADS1115 A/D converter module:......................... 21-3
Analog input: 4-2
Analog switch chip:... 3-31
Analog-to-digital converter module: 3-27
Anderson Powerpole connectors: 13-5
Android apps: 26-2
Arduino
    History:.. 1-3
    Uno: 1-2
Arduino boards and variants:............................. 2-1ff
    Bluetooth:.. 2-2
    Comparison chart:... 2-12
    DC Boarduino:.. 2-4
    Diecimilia:... 2-2
    Digilent chipKIT Max32:................................ 2-11
    Digilent chipKIT Uno32:............................... 2-10
    Due: 2-8
    Duemilanove:.. 2-2
    Esplora:.. 2-7
    Extreme:... 2-2
    Iduino Nano:.. 2-2, 2-3
    Leonardo:... 2-7
    LilyPad:... 2-2, 2-4
    Mega: 2-6
    Mini: 2-2
    NG: 2-2
    Pro Mini:... 2-3
    Solarbotics Ardweeny: 2-4
    Tre: 2-9
    Uno R3:.. 2-3
    USB: 2-2
    Yun: 2-8
Arduino Integrated Development Environment
    (IDE):.. 5-1ff
Argent Radio Data shield: 3-8
ATmega1210: 2-6
ATmega168: 2-2
ATmega328: 2-3
ATmega32u4: 2-7
ATmega8: 2-2
Audio shield:... 3-4
Azimuth/Elevation rotator controller project: 21-1ff

B

Bluetooth module: 3-25
BMP085 barometric pressure sensor: 3-21, 11-3
Breadboard shield:..................................... 3-10, 6-2

C

CDE/Hy-Gain rotator controller
    commands (Table):....................................... 25-9
CDE/Hy-Gain rotator controller
    modification project: 25-1ff
CDE/Hy-Gain rotator controller project: 24-1ff
Code practice generator project:....................... 7-1ff
Color TFT display: 3-3, 3-16
Creative Commons License:................................ 1-6
Current sensor module:..................................... 3-24
CW beacon and foxhunt
    keyer project:.. 8-1ff
CW decoder project: 22-1ff
CW iambic keyer project: 17-1ff
CW keyboard project:....................................... 19-1ff

D

DC Boarduino:.. 2-4
Debugging:.. 5-9, 21-7
Development station:... 6-2
DFRobot Graphic LCD4884 display shield:............ 3-2
dht library:... 11-5
Digital compass module:.................................... 3-24
Digital compass project:................................... 10-1ff
Digital I/O:... 4-1
Digital I/O expander: 3-30
Digital potentiometer chip:................................. 3-31
Digital-to-analog converter module:................... 3-29
Digilent chipKIT Max32:..................................... 2-11
Digilent chipKIT Uno32:..................................... 2-10
Direct digital synthesizer (DDS) module: 3-21
    AD98xx series:.. 3-21
Display:.. 3-12ff
    Color TFT: 3-3, 3-16
    DFRobot Graphic LCD4884:............................ 3-2
    Graphic LCD:... 3-2
    Hitachi HD44780:................................... 3-2, 3-12
    LCD: 3-2
    LED: 3-32

LED driver: 3-32
Nokia 5110: 3-2, 3-14, 11-3
Organic LED (OLED): 3-15
Vacuum fluorescent display (VFD): 3-13
VGA module (4D Systems): 3-16
DS18B20 temperature sensor: 3-19, 9-3

E

EaglePCB software: 5-7, 7-8, A-5
EasyVR shield: 26-2
EEPROM library: 19-5
EEPROM module: 3-30
Elliott, Steven, K1EL: 17-2
Emic 2 text-to-speech module: 3-23, 15-2, 16-5
Enclosures: 3-34ff
Ethernet shield: 3-5

F

Fan speed controller project: 9-1ff
Flow chart: 5-6
Franklin AS3935 lightning detector: 23-2
Franklin AS3935 lightning detector module: 3-24
Fritzing software: 5-7, 7-3, A-5
FTDI USB interface: 2-4, 3-18, 20-4

G

GNU GPL (General Public License): 1-5
GPS (Global Positioning System): 16-3
GPS logger shield: 3-8
GPS module: 3-23
Graphic LCD shield: 3-2

H

H-bridge chip: 3-32
Ham Radio Deluxe (*HRD*) software: 20-11, 21-1, 21-31, 24-25, 25-2, 25-21
HAM series rotator: 24-1
Hitachi HD44780 display: 3-2, 3-12
HMC5883L digital compass module: 3-24, 10-1
HMC5883L library: 10-3
Hy-Gain rotator controller modification project: ... 25-1ff
Hy-Gain rotator controller project: 24-1ff

I

I/O methods: 4-1ff
   Analog input: 4-2
   Digital I/O: 4-1
   Inter-Integrated Circuit (I2C) bus: 4-4
   Interrupts: 4-5
   MaxDetect 1-Wire interface: 4-3
   Maxim 1-Wire interface: 4-2
   Pulse width modulation (PWM): 4-2
   Serial I/O: 4-2
   Serial Peripheral Interface (SPI) bus: 4-3
I/O shield: 3-10
Iambic keyer modes: 17-1
Iduino Nano: 2-3
INA169 current sensor module: 3-24, 13-2
Inches of mercury: 11-2
Inkscape software: A-5

Integrated Development Environment (IDE): 5-1ff
Inter-Integrated Circuit (I2C) bus: 4-4
Interrupts: 4-5

L

LCD display: 3-2, 3-12
LCD5110_Basic library: 11-5, 16-12
LED display: 3-32
LED driver: 3-32
Lesser GNU GPL (General Public License): 1-6
Level converter module: 3-18
Library: 5-3ff, A-1, A-4
   dht: 11-5
   EEPROM: 19-5
   HMC5883L: 10-3
   LCD5110_Basic: 11-5, 16-12
   LiquidCrystal_I2C: 7-5, 10-3
   Math: 16-7
   Morse: 7-6, 8-4
   MorseEnDecoder: 22-4
   OneWire: 9-5
   PS2Keyboard: 19-3
   SoftwareSerial: 15-5, 16-8
   TinyGPS: 16-5
License, Open Source: 1-4ff
Lightning detector module: 3-24
Lightning detector project: 23-1ff
LilyPad Arduino: 2-2, 2-4
LiquidCrystal_I2C library: 7-5, 10-3
LM567 tone decoder: 22-2

M

Maidenhead grid locators: 16-2
Math library: 16-7
MAX7219 LED driver: 3-32, 12-2
MaxDetect 1-Wire interface: 4-3, 11-2
Maxim 1-Wire interface: 4-2, 9-3
MCP4725 D/A module: 18-4
Memory management: 5-8
Memory tracking: 19-5
Millibars: 11-2
MIT License: 1-6
MOD-1016 lightning detector module: 23-4
Module: 3-11ff
   Analog-to-digital converter: 3-27
   Bluetooth: 3-25
   BMP085 barometric pressure sensor: 3-21, 11-3
   Current sensor: 3-24
   Digital compass: 3-24
   Digital-to-analog converter: 3-29
   Direct digital synthesizer (DDS): 3-21
   DS18B20 temperature sensor: 3-19, 9-3
   EEPROM: 3-30
   Emic 2 text-to-speech: 3-23
   Franklin AS3935 lightning detector: 3-24
   GPS: 3-23
   HMC5883L digital compass: 3-24
   INA169 current sensor: 3-24
   Level converter: 3-18
   Motion detector: 3-32
   Motor driver: 3-27
   Real-time clock/calendar (RTCC): 3-26

RHT03 humidity/temperature sensor: 3-19, 11-2
SD card: 3-27
Skylab SKM53 GPS: 3-23
Text-to-speech: 3-23
TinyRTC clock/calendar: 3-26
Vibration sensor: 3-32
Morse library: 7-6, 8-4
MorseEnDecoder library: 22-4
Motion detector module: 3-32
Motor driver module: 3-27
Motor driver shield: 3-4
Multimeter: 6-3

N

NMEA protocol: 16-1, 16-4
Nokia 5110 display: 3-2, 3-14, 11-3

O

On-air indicator project: 14-1ff
OneWire library: 9-5
Open Source: 1-4ff
    Creative Commons License: 1-6
    GNU GPL (General Public License): 1-5
    Lesser GNU GPL: 1-6
    MIT License: 1-6
Organic LED (OLED) display: 3-15
Oscilloscope: 6-4

P

Pascals: 11-2
Prototyping shield (protoshield): 3-11, 6-2
PS2Keyboard library: 19-3
Pulse width modulation (PWM): 4-2

R

Real-time clock/calendar (RTCC) module: 3-26
Relay shield: 3-3
Resistor to resistor ladder network: 18-2
RF probe with LED bar graph project: 12-1ff
RHT03 humidity/temperature sensor: 3-19, 11-2

S

Satellite tracker project: 20-1ff
SatPC32 software: 20-2, 21-1, 21-31
Schematic diagram: 5-6
SD card module: 3-27
SD card shield: 3-7
Sensor
    BMP085 barometric pressure sensor: 3-21, 11-3
    DS18B20 temperature sensor: 3-19, 9-3
    RHT03 humidity/temperature sensor: 3-19, 11-2
    Vibration sensor: 3-32
Serial I/O: 4-2
Serial Monitor: 21-7
Serial Peripheral Interface (SPI) bus: 4-3
Servo: 20-2
Shield: 1-2, 3-1ff
    Argent Radio Data: 3-8
    Audio: 3-4
    Breadboard: 3-10, 6-2

Color TFT display: 3-3
DFRobot Graphic LCD4884 display: 3-2
EasyVR: 26-2
Ethernet: 3-5
GPS logger: 3-8
Graphic LCD display: 3-2
Hitachi HD44780 display: 3-2
I/O: 3-10
LCD display: 3-2
Motor driver: 3-4
Nokia 5110 display: 3-2
Prototyping (protoshield): 3-11, 6-2
Relay: 3-3
SD card: 3-7
USB Host: 3-6
WiFi: 3-6
Xbee: 3-8
Skylab SKM53 GPS module: 3-23, 16-3
SoftwareSerial library: 15-5, 16-8
Solar battery charge monitor project: 13-1ff
Solarbotics Ardweeny: 2-4, 20-3
Soldering tools: 6-3
Switches: 3-32
SWR sense head: 15-3

T

Talking GPS/UTC time. grid square indicator
    project: 16-1ff
Talking SWR meter project: 15-1ff
TEN-TEC Rebel Model 506 transceiver: 26-3
Test equipment: 6-3
Text-to-speech module: 3-23
TinyGPS library: 16-5
TinyRTC clock/calendar module: 3-26

U

USB Host shield: 3-6

V

Vacuum fluorescent display (VFD): 3-13
Vendor links: A-7
VGA display: 3-16
Voice recognition: 26-2

W

Waveform generator project: 18-1ff
Weather sensors: 3-19ff
Weather station project: 11-1ff
WiFi shield: 3-6

X

Xbee shield: 3-8

Y

Yaesu G5400/5500 rotator: 21-1
Yaesu GS-232A rotator controller
    commands (Table): 21-5
Yaesu GS-232A rotator interface: 20-2, 21-1, 25-21